Praise for *Why We Lie*

"Self-deception is one of the most powerful ideas in psychology, indeed, in human affairs, and David Smith's *Why We Lie* is an excellent synthesis of this crucial topic. The biology is up-to-date and accurate, the psychological implications are clearly worked out, and the writing is inviting and accessible."

—Steven Pinker, bestselling author of
The Blank Slate and *The Language Instinct*

"David Smith has pulled off a beaut. *Why We Lie* is a wonderfully blended cluster of arguments to support the painful truth that we are a species whose skills at deceiving others is matched only by our ability to deceive ourselves."

—Arthur S. Reber, author of *The Penguin Dictionary of
Psychology* and *The New Gambler's Bible*

"*Why We Lie* is a fascinating book about a fascinating subject. . . . rich with stories, anecdotes, and psychological as well as sociological analyses."

—Tamar Frankel, S.J.D., *The Human Nature Review*

Why We **LIE**

The Evolutionary Roots
of Deception and the
Unconscious Mind

David Livingstone Smith

St. Martin's Griffin
New York

www.stmartins.com

Library of Congress Cataloging-in-Publication Data

Smith, David Livingstone, 1953–.
    Why we lie: the evolutionary roots of deception and the unconscious mind / David Livingstone Smith.
        p. cm.
    Includes bibliographical references (p. 207).
    ISBN-13: 978-0-312-31040-0
    ISBN-10: 0-312-31040-4
    1. Deception. 2. Self-deception. 3. Subconsciousness. 4. Sociobiology. I. Title.

BF637.D42 S65 2004
153.6—dc22   2004040961

D   10 9 8 7 6 5 4 3 2

In memory of my grandparents,

Herman and Bertha

Contents

Acknowledgments

This volume was a long time coming, and would never have seen the light of day if not for the determination and support of my agent, Michael Psaltis. I was also extremely fortunate to have as my editor Ethan Friedman, who provided just the right mixture of criticism and support. Thanks also to David Buss, Steven Pinker, Arthur Reber, and Robert Trivers for their generous support and to Howard Bloom for his one-man chorus of approval. Thanks also to Leif Ottesen Kennair, Nina Strohminger, Irwin Silverman, Marilyn Taylor, Norrie Feinblatt, Kenneth J. Silver, and Felicia Sinusas.

I owe an incalculable debt of gratitude to Rob Haskell for his endless support, encouragement and stimulation, and sterling integrity. The many hours we have spent discussing unconscious communication and reassuring one another of our mutual sanity have been crucial to this book. Discussions with Steven Kercel about the mathematics of self-reference, impredicative logic, and semantic ambiguity have also had a significant impact on my thinking.

Finally, I could not have managed without the support of my wonderful wife, Subrena. Not only because of her profound intuitive understanding of evolutionary biology, which contributed several important insights, but also for her forbearance while yet again I spent a frantic summer strung out on black coffee as I struggled with an overdue manuscript.

Biology whispers deep within us.

—David Barash

*To tell deliberate lies while genuinely believing in them,
to forget any fact that has become inconvenient, and then,
when it becomes necessary again, to draw it back from
oblivion for just so long as it is needed, to deny the existence of
objective reality and all the while to take account of the reality
which one denies—all this is indispensably necessary.*

—George Orwell

Why We **LIE**

Preface

At every level, from brute camouflage to poetic vision, the linguistic capacity to conceal, misinform, leave ambiguous, hypothesize, invent, is indispensable to the equilibrium of human consciousness.

—George Steiner

The human mind is one of the most extraordinary and most poorly understood characters in Mother Nature's great gallery of creations. It took millions of years for it to evolve. Over immense expanses of prehistory, our ancestors acquired minds with a distinctively human cast—a range of passions and emotions, the ability to express their thoughts in words, to craft tools, to plan and to lie. Unfortunately or not, the gradual changes in brain structure that eventually produced the modern mind did not endow us with much ability to understand ourselves. Self-understanding does not come naturally to human beings, like eating, drinking, and having sex. Pursuing the reasons for this takes us to the heart of human nature.

Evolutionary biology teaches us that the tendency to deceive has an ancient pedigree. We find it in many forms, at all levels, throughout the natural kingdom. Even viruses, organisms so simple that it is a struggle to think of them as living things, have subtle strategies for deceiving the immune systems of their hosts: nature is awash with deceit. Many of the phenomena that I am going to describe would have seemed wildly implausible in the era before biologists got down to the disciplined, scientific observation of animal behavior. In fact, even today there are people who have neither direct experience observing animals nor knowledge of the scientific literature who are prone to greet this sort of material with statements such as "Are you putting me on?" The moral of this story is to caution you against hastily rejecting some of the claims that I will make about human deception just because they seem far-fetched. Nature *is* far-fetched. Deceptive creatures have an edge over their competitors in the relentless struggle to survive and reproduce that drives the engine of evolution. As well-honed survival machines, human beings are also naturally deceptive.

Deceit is the Cinderella of human nature; essential to our humanity but disowned by its perpetrators at every turn. It is normal, natural, and pervasive. It is not, as popular opinion would have it, reducible to mental illness or moral failure. Human society is a "network of lies and deceptions"[1] that would collapse under the weight of too much honesty. From the fairy tales our parents told us to the propaganda our governments feed us, human beings spend their lives surrounded by pretense.

Seven million years ago our ancestors were intelligent apes that lived in complex social groups dominated by linear hierarchies. The social world presented them with formidable intellectual challenges, and the need to cope with these exigencies pushed primates down the long road that eventually led to the evolution of anatomically modern human beings. Sheer social complexity compelled our prehuman ancestors to become progressively more intelligent, and as they did so they also became increasingly adept at social gamesmanship; the wheeling, dealing, bluffing, and conniving that I call "social poker." Once established, the need to cope with skilled social players became a selection pressure that escalated cognitive development even further.

Between about five and seven million years ago the hominid and chimpanzee lines went their separate ways. The earliest hominids were small, hairy, and did not look much different from modern apes. Over the millennia, several species of human beings came into existence and passed away into extinction; the last of these, the Neanderthals, died out only about thirty thousand years ago. Then, sometime between one hundred and one hundred fifty thousand years ago, our own kind, *Homo sapiens* (Wise Men) arrived. At some point, we do not know exactly when, our prehistoric ancestors learned to speak. This momentous step altered human society and the human mind forever. It was probably only after spoken language arrived that Wise Men became able to lie to themselves.

Why did self-deception take root in the human mind? As we will see, the propensity for self-deception probably became part of our nature because it was so helpful to us in our dealings with one another. Not only does lying to oneself soothe many of the stresses of life, but, more importantly, it also helps one lie to others. One of the most important insights of modern sociobiology is that self-deception is the handmaiden of deceit: in hiding the truth from ourselves, we are able to hide it more fully from others. Therefore, like deceit, self-deception lies at the core of our humanity. Far from being a sign of emotional disturbance, as both popular and psychiatric folklore suggest, it is probably vital for psychological equilibrium. The first aim of this book is to make the details of this evolutionary theory of self-deception available to a wider audience.

The ability to harness the magic of words, and the capacity for self-deception that came it its wake, reconfigured the human psyche. In order to hide the truth about ourselves from ourselves, we needed to evolve an unconscious mind. Evolutionary biology implies that there is a region of our mind devoted to our dealings with other people that never divulges its secrets to conscious awareness. There is a side of ourselves that we were evolved not to know.

The great psychologist and philosopher William James cautioned over a century ago that talk of the unconscious "is the sovereign means for believing what one likes in psychology, and of turning what might become a science into a tumbling-ground for whimsies."[2] We need to tread carefully, but due caution is not the same as intellectual meekness and lack of imagination. I have tried to hug the scientific coastline as much as possible, while at the same time doing justice to the phenomena. I have made a particularly strenuous effort to avoid the all-too-common tendency to deny the existence of something just because it does not fit into the procrustean bed of existing knowledge. Notwithstanding these efforts, I am sure that some readers will regard this book—particularly its later chapters—as unacceptably reckless.

As soon as we broach the topic of the unconscious mind, we must pay our respects to Sigmund Freud. Freud's work was critical for establishing the idea of the unconscious on our intellectual landscape. It is certain, though, that many of Freud's ideas about the unconscious mind were far off the mark. Since the 1950s, experimental psychologists have developed ideas about unconscious mental processing that are very different from the Freudian ones, and nowadays, cognitive scientists propose that most of our mental processes happen outside of awareness. However, something has been lost in this transformation. In their rush to capture the quicksilver of mind in the nets of laboratory methods and computational models, psychologists have neglected the messy domain of "hot" cognition, of passion and conflict. Self-deception does not lend itself to experimental investigation, and consequently empirical psychologists have had little to say about it. A second aim of this book is to reconnect cognitive psychology with the kinds of questions that Freud tried unsuccessfully to answer and which evolutionary biology has reintroduced to the scientific agenda.

"There seems to be little doubt," wrote the biologist David Barash, "that the unconscious, although poorly understood, is real and that in certain obscure ways it influences our behavior. We can, therefore, predict that it is a product of our evolution, and, especially insofar as it is widespread and 'normal,' that it should be an adaptive product as well."[3] The third task of this book is to describe some of the adaptive functions of the unconscious mind implied by an important but widely overlooked implication of the main evolutionary theory of self-deception. I will argue that unconscious deceivers must also be unconscious perceivers; it is not possible for a person to hoodwink another without keenly observing and interpreting the reactions of the other from one moment to the next. So, if the standard evolutionary theory of self-deception is correct, if we deceive ourselves in order to deceive others, unconsciously we must all be natural psychologists, carefully monitoring one another's behavior and drawing subtle inferences about each other's mental states *without having the slightest idea that we are doing this*. We are all, borrowing a term from biologists John Krebs and Richard Dawkins, unconscious "mind-readers."[4]

The ability to analyze unconsciously the meaning of the behavior of people around us is an essential aspect of our evolved social intelligence. As we will see, there are several competing conceptions of the "unconscious mind" described in the psychological literature, and there is good evidence that very many, if not all, mental processes are at their core unconscious. The concept of the unconscious mind that I present in this book is quite different from either the Freudian id, a seething cauldron of untamed and irrational urges, or the coolly mechanical neurocomputational processes nowadays called the cognitive unconscious. The social unconscious must be smart, in fact very smart, as well as highly adaptive. It must also be—again, for reasons that will emerge later in this book—sequestered from our conscious social perceptions and judgements. I call it the "Machiavellian module" to distinguish it from the Freudian and standard cognitivist models.

Because there has been little empirical research into the unconscious dynamics of social relations, we need to explore avenues that cognitive scientists usually neglect, beginning with the writings of Sigmund Freud. Unfortunately, any mention of Freud is likely to raise eyebrows among members of the psychological fraternity unless it is purely dismissive. However, this prejudice need not deter us. There are several tantalizing remarks buried in Freud's writings that point to an immensely sophisticated, yet completely unconscious, module for social intelligence. Moreover, just before his death in 1932, Freud's Hungarian colleague Sándor Ferenczi reported observations that seem to show this mental module in action. Sketchy and impressionistic though they are, the remarks by Freud and Ferenczi provide valuable clues for understanding how the human mind works.

Ferenczi's comments suggest, somewhat surprisingly, that raw, unconscious perception influences how we communicate with one another. He thought that unconscious thoughts are expressed in something like a verbal code, and that the narrative images found in ordinary conversations can manifestly convey one set of meanings while covertly expressing quite another. This makes biological sense. The deceptions and manipulations that we impose on one another must be concealed behind a shroud of secrecy in order to work, a shroud that is so opaque that even the perpetrator remains largely unaware of his or her own maneuvers. The idea that we are incorrigibly aware of the full meaning of what we say is a hangover from the antiquated, Cartesian view that the mind is transparent.

These considerations open an extremely important, but all

too often neglected, aspect of human interaction. Researchers on the evolution of language have so far been almost entirely interested in the *literal* use of words, as though hominids have always called a spade a spade. However, if one of the basic functions of language is to deceive others, a good evolutionary theory should have something to say about its intrinsically ambiguous, nonliteral dimensions. Euphemisms, puns, double entendres, and other kinds of coded speech use metaphor and analogy to express meanings indirectly. Unlike many psychologists, literary scholars are well aware of the centrality of such wordplay for human speech. When George Steiner linked "poetic vision" with "the linguistic capacity to conceal" in the quotation at the beginning of this preface, he was right. The fourth, and by far the most controversial, aim of this book is to argue that many of our unconscious perceptions of one another are disguised in the stories that we tell in everyday conversation. Human beings are compulsive narrators. We spend the bulk of our conversation time telling stories, usually of the kind derogatively called gossip. Information self-deceptively excluded from consciousness seeps between the semantic cracks of ordinary speech and much of our everyday chitchat has a powerful, figurative dimension that portrays the unvarnished truth concealed behind the genteel façade of ordinary social interaction. The unconscious poetry of everyday conversation voices thoughts that we bar from conscious awareness. I know that this will sound outrageous to some readers. Surely, they will object, you cannot mix up science with poetry! My response is not only that we can, but also that we *must* if we are to understand human behavior, for poetry, too, is part of nature.

Ludwig Wittgenstein distinguished between two varieties of originality: that belonging to the seed and that belonging to the soil.[5] I wish that I could lay claim to originality of the first kind. The seeds that grew into this book were sown by many fine scholars. I hope that I do not deceive myself in thinking that it is the originality of the soil, the way that I have drawn out implications and made connections between ideas that might, at first glance, seem completely unrelated to one another, that makes this book worth reading.

A book purporting to be scientific, but which builds arguments on what may seem to them to be the flimsiest of foundations, is bound to disappoint some readers. Where is the experimental data? I plead guilty of not having provided adequate empirical support for the distinctive views advanced in this book. Although immensely powerful and valuable, experimental research is not the be-all and end-all of cognitive science. Science is the disciplined passion to find out, to make sense of the puzzles presented by the world around us and within us, and to make reasoned extrapolations from what we already know, or at least think that we know. This book points the way to empirical research into self-deception and unconscious communication. It shows scientists of the mind where they might look, and just what phenomena they ought to look for.

You might say that this book resides on the borderland between science and science-yet-to-be, in a wild place where there are few paths and the signposts are scarce and difficult to read. I hope that you enjoy the journey.

1

Natural-Born Liars

Lying is universal—we all do it; we all must do it.

—Mark Twain

Mel dug furiously with her bare hands to extract the large suc-
culent corm from the rock-hard Ethiopian ground. It was the
dry season and food was scarce. Corms are edible bulbs rather
like onions, and are a staple during these long hard months. Little Paul sat nearby and surreptitiously observed Mel's labors out
of the corner of his eye. Paul's mother was out of sight. She had
left him to play in the long grass; but, he was secure in the
knowledge that she would remain within earshot in case he
needed her. Anyway, at this moment he was concerned with Mel
rather than with the precise whereabouts of his mother. Just as
Mel managed, with a final heave, to yank her prize out of the
earth, Paul's ear-splitting cry shattered the peace of the savanna.
His mother rushed to her boy. Heart pounding and adrenaline
pumping, she burst on the scene and quickly sized up the situation: Mel had obviously harassed her darling child. Furiously
shrieking abuse, she stormed after the bewildered Mel, who
dropped the corm and fled. Now Paul's scheme was complete.
After a furtive glance to make sure nobody was looking, he
picked up his prize and began to eat. The trick worked so well
that he used it several more times before anyone wised up.

Kids will be kids, even when they are apes. The anecdote

that I have just recounted describes the behavior of a juvenile chacma baboon observed by the primatologist Richard Byrne.[1] It illustrates the fact, which has long been known to biologists, but has more recently been shown to have enormous consequences for our conception of the human mind, that the roots of deceit lie deep in our biological past. Although in many ways impressive, the social manipulations of baboons, chimpanzees, and other non-human species are easily finessed by our own talent for deceit. Human beings are grandmasters of mendacity. It would not have been out of place to name our species *Homo fallax,* (deceptive man), instead of *Homo sapiens* (wise man). To understand why, we need to explore the origins of the modern human mind.

The Stone Age Mind

Darwin predicted in *The Origin of Species* (1859) that evolutionary theory would one day provide a new foundation for the science of psychology; but it would be more than a century until the truth in his words was borne out. The change came when advances in our understanding of the genetics of social behavior ushered in the controversial new science of sociobiology, the biological study of the social behavior of humans and other animals.[2] Before the pioneering work of the Harvard biologist Edmond O. Wilson, the study of human social behavior had been dominated by the dogma of cultural determinism. According to this view, which remains prevalent in the social and behavioral sciences, the forces of culture are all-powerful in shaping human behavior. Culture itself is said to be autonomous, standing outside of and relatively untouched by the primitive forces of nature. Primed by the still fresh memory of Nazi eugenics, the effort to "purify" the human race by killing or sterilizing "defectives," many social scientists were deeply suspicious of any theory purporting to describe the biological foundations of human nature. Some of them luridly portrayed sociobiologists as dangerous neo-Fascists, hell-bent on racism, sexism, and the preservation of the political status quo.[3] Over the next three decades, human sociobiology transformed itself into evolutionary psychology, an approach to psychological science that studies the mind from the standpoint of its prehistoric and evolutionary origins. Evolutionary psychology is not just one more school of psychology. It is a perspective on the *whole* of psychology that claims that we are human animals, and that our minds, no less than our bodies, are products of the forces of nature operating on a time frame of millions of years; human nature was forged from our ancestors' struggle to survive and reproduce. It is difficult to comprehend this expanse of time without some help. Consider it this way: if all the time that has elapsed since the emergence of the first hominids were a single day, the whole period of recorded history, some five thousand years, would occupy only the *final two minutes*.

Remains of prehistoric skulls suggest that the human brain attained its present form about one hundred fifty thousand years ago. We lived in an environment very different from that of all but a very few human populations today, eventually emerging from prehistory equipped with an array of passions, skills, and mental abilities specifically adapted to life in that primeval habitat. The mind that you and I possess is, in its essentials, a Stone Age mind.

Evolutionary biology does not endorse the popular and reassuring conviction that human minds are tools for self-knowledge and the pursuit of truth. The human mind evolved for the very same reason that all of our other organs evolved; namely, because it contributed to its owners' reproductive success. Nature selected those mental capacities that helped to spread our genes, and those that proved unhelpful were ineluctably snuffed out. As any seducer knows, honesty and reproductive success are not necessarily good bedfellows. Because deception and self-deception helped our species to succeed in the never-ending struggle for survival, natural selection made them part of our nature. We are deceptive animals because of the advantages that dishonesty reaped for our ancestors, and which it continues to secure for us today. But I am getting ahead of myself. Let me first survey the landscape of human deceit and leave the discussion of its evolution for subsequent chapters.

The Ubiquity of Deceit

Deceit is and probably always has been a major concern of human culture. The founding myth of the Judeo-Christian tradition, the story of Adam and Eve, revolves around a lie. We have been talking, writing, and singing about deception ever since Eve told God "The serpent deceived me, and I ate." Our seemingly insatiable appetite for stories of deception spans the extremes of culture from *King Lear* to Little Red Riding Hood. These tales are so enthralling because they speak to something fundamental in the human condition. Deception is a crucial dimension of all human associations, lurking in the background of relationships between parents and children, husbands and wives, employers and employees, professionals and their patients, governments and their citizens.

Lying is obliged by its very nature to cover its traces, for in order to lie effectively we must lie about lying. This poses a problem for anyone attempting to prove the ubiquity of deception. Although it is all around us, deception is strangely elusive, "hard to explain, although it is something with which we are all intimately familiar."[4] We do not need the surveys and experiments beloved by psychologists to confirm that people often lie to each other, although these, too, have proven to be quite revealing. To grapple with dishonesty, we have to open our eyes to some unpleasant truths. As the biologist William Hamilton once remarked, evolutionary thinking about human behavior is not difficult in the way that doing physics is. It does not require highly sophisticated mathematics, elaborate instrumentation, or difficult chains of logic. Viewing human behavior through a Darwinian lens is difficult because it radically undermines cherished illusions about human nature. It leads us to violate mental taboos, to enter no-go areas, to open the book of forbidden knowledge. It is "socially unthinkable," exposing the raw nerves of our relationships with one another and revealing the complex manipulative strategies that oil the wheels of society.[5] Thinking biologically about human nature means dismantling shared illusions.

Although we claim to value truth above all else, we are also at least dimly aware that there is something antisocial about too much honesty. This dilemma has often been portrayed in literature and film, from Dostoevsky's Prince Mishkin, whose innocence and honesty destroy the lives of those around him, to the 1997 film *Liar, Liar!* in which a lawyer wreaks havoc when he is placed under a spell condemning him to be truthful for twenty-four agonizing hours. Evolutionary biology suggests that no normal person would be capable of such a feat. We are natural-born liars.

What is a Lie?

When we think of lying, we typically think of explicit verbal falsehoods. The philosopher Sissela Bok, who is a spokesperson for this view, defines a lie as any intentionally deceptive statement.[6] Is this all there is to lying? Mark Twain didn't think so, and reckoned that "by examination and mathematical computation I find that the proportion of the spoken lie to the other

varieties is 1 to 22,894. Therefore the spoken lie is of no consequence, and it is not worthwhile to go around fussing about it
and trying to make believe that it is an important matter."[7] My
own sympathies lie with Twain rather than with Bok, because
Twain's perspective is both inclusive and biologically realistic.
As we saw with Paul and Mel, and we will see in chapter 2, deception is not the exclusive province of our species. Many other
organisms make liberal use of deception to get their way. I
therefore define lying as *any* form of behavior the function of
which is to provide others with false information or to deprive
them of true information.

I purposefully use the term "function" rather than "intention." In the vocabulary of evolutionary biology, the function of
something is that which it has been selected (metaphorically,
"designed") to do. Consider the bodies of leaf insects, which
mimic the form and color of the plants that they inhabit. These
bugs do not *intend* to deceive the creatures that want to make a
meal of them and can no more change their physical shape
than you or I can. Camouflage, a form of deception, is nonetheless a function of their bodily form. Lying, in Bok's restricted
sense of the word, is but one small detail of Twain's vast and
intricate tapestry of guile.[8]

Lying can be conscious or unconscious, verbal or nonverbal,
stated or unstated. Appreciating this is vital for any comprehen-
sive understanding of deceit, and is perhaps the most important point raised in this chapter. Think for a moment of all the
forms of dishonesty that do not require the use of explicit verbal falsehoods. Breast implants, hairpieces, feigned illnesses,
faked orgasms, and phony smiles are just a few examples of
nonverbal lying. Consider also the cunning use of innuendo,
strategic ambiguity, and crucial omission, as epitomized by Bill
Clinton's infamous declaration that he "did not have sexual
relations with that woman . . . Ms. Lewinsky."

According to the folklore of deception, ordinary, decent people lie only occasionally and inconsequentially except in extreme, morally defensible circumstances. Anything more than the occasional white lie is considered a symptom of madness or badness: the penchant of the mentally ill, criminals, lawyers, and politicians. Good liars, so the myth goes, always know what they are doing: they are calculating and exquisitely aware of their deceptions. People who lie without knowing that they are lying are thought to be at best confused and at worst insane. Evolutionary psychology opposes this cozy mythology. Lying is not exceptional; it is normal, and more often spontaneous and unconscious than cynical and coldly analytical. Our minds and bodies secrete deceit.

"Everybody," writes science writer and television producer Sanjida O'Connell, "lies regularly."

> *Undergraduates lie to their mothers in half of their conversations and to complete strangers eighty per cent of the time . . . usually for financial gain (there is one price for books in the bookshop and another when parents ask), to make their friends feel better about themselves and to con their family into thinking they were studying and not in the pub the night before. People tell fewer lies to those who are close to them, but partners are likely to be lied to a third of the time, which is more than people lie to their best friends.[9]*

The everyday game of strategic impression management seethes with deception. In fact, we take our mendacity so completely for granted that we rarely reflect on it. Pause, look, think, and you will emerge with an enhanced appreciation of the enormous range of human dishonesty. University of Massachusetts psychologist Robert Feldman had filmed ten-minute "get to know me" conversations between student volunteers and a stranger and later had his subjects view the tapes to count the number of lies that they told. He found that on average, people tell three lies for every ten minutes of conversation.[10] This sounds like quite a lot of lying, but considering the fact that his subjects were unlikely to have been entirely truthful with him or with themselves, and also bearing in mind that this research measured only the frequency of narrow, explicit verbal lying, the real rate of deception must be considerably higher.

Our very appearance is often carefully arranged to present a not-totally truthful image of ourselves to the world. "Why," inquires the sociobiologist Richard Alexander, "if the truth is our goal and motto, do we begin to deceive from the moment we arise from our beds in the morning, with clothes that modify our body shapes flatteringly, makeup and hair arrangements that improve our eyelashes or face shapes or cover a bald spot?"

Why do we spend our waking hours before and after sleep, and while shaving . . . or showering or dressing, building scenarios by which we may deceive or best in some fashion those with whom we are scheduled to interact during the day? Why do we exclaim enthusiastically upon meeting someone we would rather have avoided . . . ? Why do we constantly deceive everyone?[11]

Clothes can magically transform a body by artfully manipulating viewers' attention. Padded shoulders, beloved of "power dressers," create the illusion of size and intimidating strength. Conversely, clothing can say "I am not a threat" by emphasizing conformity. High-heeled shoes, push-up bras, and apparel that exaggerate the contrast between hips, waist, and bosom create the illusion of hypersexuality. Decking oneself out in black gives the impression of slenderness, while bright colors, patterns, or accessories draw the eye away from an unflattering feature or toward a physical asset. Skillful visual prestidigitation can make a plump waist slender, a large bottom small, or small breasts prominent. Clothing and jewelry are also used to create an image of wealth, and hence, of desirability.

Much the same can be said of blatantly deceptive cosmetic devices such as hairpieces and dyes, age-concealing makeup, and hair removal, all of which are used to falsely suggest the promise and excitement of youth. The magic of cosmetics and dress is no modern innovation. Archeological evidence reveals that at least 70,000 years ago our ancestors adorned their bodies with ochre, a red powder which has been used continuously for that purpose ever since. Pleistocene women may have initially used this forerunner of rouge to deceive males by simulating menstruation, or perhaps to simulate the subtle blush that accompanies ovulation.[12] Ancient Egyptian women applied a green paste, rather like modern-day eye shadow, to define their features, darkened their eyebrows with kohl, colored their eyelids to make their eyes appear larger, and bedecked their heads with elaborate wigs. Their Mesopotamian sisters adorned themselves with paint to exaggerate the color and fullness of their lips, while fashionable Greek and Roman ladies dyed their hair blond, applied makeup to cover blemishes, lightened their skin, and used pumice to remove unwanted body hair.[13]

Moving from the universe of vision to that of smell, we apply deodorants to disguise our scent and douse ourselves in perfume to create alternative body odors. Ancient Egyptian women attended dinner parties with scented wax cones on their heads. The sultry heat of the Egyptian evening caused the melting wax to drench their wigs in exotic perfumes imported for this purpose from sub-Saharan Africa. Cleopatra, who authored a book on the art of scent making, received her future lover Mark Anthony dressed as Aphrodite, the goddess of love, on a barge equipped with fragrance-drenched sails. Ancient perfuming practices were incredibly elaborate compared to those of the present day. A Greek woman would not simply splash on some perfume before popping down to the agora. The poet Antiphanes informs us that she anointed her feet with

Egyptian scent, her cheeks and nipples with palm oil, one of her arms with bergamot, her eyebrows and hair with marjoram, and her knee and neck with thyme, all designed to deceive the male as to her true scent.[14]

Deception follows us from birth to death, and seeps into every corner of our public and private lives. Most of us claim that we try to teach our children not to lie. While it is true that children are often told not to lie, they are actually more frequently taught how to lie in a socially acceptable manner. They are instructed, on pain of punishment, to feign respect for their elders, to write heartfelt thank-you notes for disappointing Christmas presents, and to refrain from telling grandma that her breath stinks. Children learn to practice those forms of deception that are publicly prohibited but covertly sanctioned. Socially appropriate lying is not merely tolerated, it is mandatory. The child who fails to master this skill pays the heavy price of disapproval, punishment, and social ostracism.

Adults also teach children deceit by example, deceiving them by "lullabies, promises, excuses, bedtime stories, and threats about dangers in the world."[15] Although told that they must never lie to their parents, these same parents do not hesitate to regale their offspring with supposedly truthful narratives of Santa Claus and the Easter Bunny. Nonverbal deception is probably part of an infant's hard-wired psychological survival kit, but explicit verbal lying is a developmental acquisition that is dependent on a high level of cognitive sophistication. Children who are unable to lie, as George Washington reputedly was (in yet another lie often told to children), are not "good" boys and girls: they may quite possibly be autistic.[16]

As we grow older, our talent for dissimulation becomes more finely honed. In one study, 92 percent of college students admitted that they had lied to a current or previous sexual partner, and the researchers ended up wondering whether the remaining

8 percent were lying.[17] One in three job applicants lies when seeking employment.[18] Once employed, they work under managers who routinely use deceit to maximize productivity.

Men are notorious for exaggerating the extent of their sexual exploits, but research shows that women's lies about erotic encounters move in the opposite direction. When psychologists Terri Fisher and Michelle Alexander asked women to fill out questionnaires on their sexual behavior and attitudes, they found that those wired up to a phony lie detector reported having had twice as many lovers as those who weren't.[19] Even in marriage, which our culture holds up as the very paragon of intimacy and trust, we find that partners keep secrets from one another about money, past experiences (particularly sexual experiences), current flirtations, "bad" habits, aspirations and worries, their children, real opinions about friends and relatives, and so on.[20] Quite a substantial number of married people in the United States either have had or are currently conducting at least one clandestine affair.[21] The sociologists Philip Blumstein and Pepper Schwartz interpret this to mean that marriage makes people more deceptive than when they are single, but it's more likely that it is just one of the many arenas where we try to have our cake and eat it too.[22]

Deception would appear to be the norm rather than the exception in business, and it is so commonplace on Madison Avenue that an advertising industry without it is hard to imagine.[23] The almost reflexive dishonesty of politicians is legendary. Not even the family doctor emerges untarnished: 87 percent of physicians surveyed felt that it is acceptable to deceive their patients in certain circumstances (including deceiving the spouse of a patient who has contracted a sexually transmitted disease in an extramarital affair).[24]

Deception is key to success in combat. Two thousand years

ago Sun Tzu wrote in his handbook for generals that "All warfare is based on deception."[25]

True excellence is to plan secretly, to move surreptitiously, to foil the enemy's intentions and balk his schemes, so that at last the day may be won without shedding a drop of blood.[26]

The story of the Trojan Horse, if not literally true, gives us a metaphor for the intimate relationship between warfare and deceit. From the Biblical yarn of Gideon and the Midianites to modern stealth aircraft, the history of warfare is largely one of collective deception and counterdeception.[27] The language of war distributed for public consumption is full of euphemisms such as "liquidation," "collateral damage," and "take out" to conceal brutal realities. The myth of the noble hero is one of the most entrenched lies of modern warfare. For much of the twentieth century, violent, mentally unstable individuals, and especially ex-convicts were seen as particularly desirable recruits. As one World War I military psychologist put it, the best soldier is "more or less a natural butcher, a man who can easily submit to the domination of intellectual inferiors."[28] Men in the field often did not have a very high opinion of "heroes," regarding them as "inhuman and unreliable."[29] On the individual level, too, a good fighter needs to know how to feint. Joe Torres, a former light heavyweight boxing champion of the world, once remarked to Floyd Patterson, "A feint is an outright lie. . . . A left hook off the jab is a classy lie."[30]

Deception is excusable in warfare, but not in science. Nevertheless, some of our most revered scientific icons have not been averse to fudging the data when it suited their purposes. Sir Isaac Newton attempted to mollify critics of his masterpiece, the *Principia Mathematica*, by modifying data to agree precisely with his theory, eventually claiming an implausible degree of precision in his measurements of physical phenomena.[31] In the year 1712, an acrimonious priority dispute broke out between Newton and the German philosopher/scientist Gottfried Leibniz over the invention of calculus. It was resolved in Newton's favor in a report published by the British Royal Society. In fact, the entire report, which tendentiously supported Newton's claim, was clandestinely penned by Sir Isaac himself![32] In 1936, the distinguished geneticist Sir Ronald Fisher pointed out that Gregor Mendel, the scientific titan who discovered the fundamental principles of genetics, appears to have "adjusted" his data in much the same manner as Newton. Mendel's results were just too good to be true.[33] Sigmund Freud, who has long been regarded as a paragon of unflinching honesty, has been devastatingly portrayed in recent scholarship as a serial deceiver.[34] The elaborate precautions taken by modern science can constrain but not obliterate the all-too-human inclination to deceive.

It is not true that lying is usually conscious and calculating. As a species, we are so well practiced in the art of deception that it comes to us almost as naturally and effortlessly as breathing. Just try keeping track of the countless deceptions that you, and all of us, perpetrate on an ordinary day and you will quickly discover that you rarely have to work at deception. Even verbal lies slide off the tongue so effortlessly that they often remain unnoticed by their fabricator, which leads us to the topic of *self*-deception.

The Puzzle of Self-Deception

Not only do we find it all too easy to deceive others, but also we are equally adept at deceiving *ourselves*. As is the case with lying, I prefer an expansive definition to a restrictive one: self-deception is any mental process or behavior the function of which is to conceal information from one's own conscious mind.

Self-deception has been a puzzle for psychologists and philosophers for more than two millennia. There seems to be something inherently paradoxical about a person simultaneously deceiving himself and being a victim of his or her own deception. The popular view of self-deception is strongly negative. Lying to oneself is supposed to be rooted in fear, guilt, or mental disorder. Some thinkers find the whole idea so preposterous that they deny that genuine self-deception exists.[35] How is it possible for both the deceiver and the deceived to be the same person? Others compare self-deception to our deception of others, suggesting that self-deception must involve a fragmentation of the personality into several interacting subminds, and that self-deception happens when one of these components succeeds in fooling the others in order to get its way.[36]

We can accept that, in spite of the apparent paradox involved, self-deception is perfectly real, and we do not have to swallow the idea of subpersonalities in order to cash it in. The main obstacle to understanding self-deception is a set of false and restrictive commonsensical beliefs about the nature of the human mind. These concepts are part of what philosophers call the Cartesian worldview, because they were most powerfully and elegantly formulated by the French polymath René Descartes early in the seventeenth century. Descartes proposed that the mind is *all consciousness*. In other words, we are immediately and automatically aware of everything going on in our heads. He also asserted that we simply cannot be mistaken about what goes on in our inner world: each of us is the sole, infallible, and unimpeachable authority on our mental states. If this is true, it means that simple introspection, the practice of looking into one's own mind, would be the only method required for self-knowledge. Descartes also promoted the idea that the mind, self, or soul is a spiritual entity standing outside the messy material realm of neurons, synapses, and neurotransmitters. This fully conscious self is autonomous, capable of free will and, in the words of a twentieth-century Cartesian, "condemned to freedom."[37]

For the better part of 250 years, Descartes's theory and its later variants dominated attempts to understand the mind. One of the people instrumental in wrecking the Cartesian monopoly was a young neurologist named Sigmund Freud. Freud was well informed about new scientific investigations into hypnosis, dreams, mental illness, and organic disorders of the brain that called the Cartesian conception of the mind into question. He realized that the mind must be identical to the brain, and that the slimy ball of nerve tissue inside our skulls is somehow responsible for the totality of our subjective mental life: our thoughts, hopes, dreams, fears, and fantasies.

Freud argued that the brain contains a number of modules, functional systems that carry out specific activities. Most controversially, he proposed that the part of the brain that thinks is entirely distinct from the part of the brain that is conscious. In other words, *all thinking is essentially unconscious,* a concept we will delve into more fully in chapter 5. In order to enter consciousness, information has to pass from the thinking part to the consciousness-producing part of the brain. This flow of information takes time and is controlled by a system of cognitive filters that determine which thoughts will enter consciousness and which will remain excluded from awareness. According to Freud, it is precisely the gap between cognition and consciousness, and the cognitive bouncer standing between them, that makes self-deception possible. In Freud's story, the unified, Cartesian self is a myth. It is nothing more than an image projected onto the screen of consciousness, pure *output,* a seductive mirage produced by a massively interconnected network of electrochemical switches in the flesh-and-blood machine that we call the brain.

Introspection does not provide insight into the workings of the neural machinery that generates it any more than the display on a computer monitor provides a picture of the processes taking place within the processing unit. Furthermore, our subjective account of ourselves is highly tendentious, because the information on which it is based has been carefully edited before being "published" as a conscious self-representation. The conscious self is fiction, a creation of the mind rather than its underpinning. The wellsprings of thought, emotion, and behavior are found in that obscure region of the mind that Freud called the unconscious.[38]

Most of Freud's contemporaries were steeped in the Cartesian tradition and regarded his conception of the mind as absurd if not repugnant. Many present-day psychologists have also found it all too easy to dismiss Freud's views on consciousness and the unconscious as nothing more than dubious, pseudoscientific speculations; and, although many of Freud's specific hypotheses about the human mind have been roundly discredited, research carried out by cognitive psychologists and neuroscientists over the last forty years has vindicated much of his general conception of the architecture of the human mind. The idea that mental processes consist of nothing more than neurophysiological states was extremely radical in 1895, but is nowadays the common position. A number of psychologists now accept that consciousness displays rather than generates information.[39] Cognitive scientists routinely speak of "non-conscious" or "automatic" mental processes, often avoiding the specific term "unconscious," which means the same thing, for fear of being tarred with the Freudian brush. The literature on experimental social psychology has repeatedly confirmed the claim that the information about our mental processes provided by introspection is highly unreliable.[40]

The notion that human beings deceive themselves about their

own desires is the leitmotif of Freudian psychology, but Freud and his followers remained blithely unconcerned about providing any scientific evidence for their claims. Such evidence would have certainly strengthened the plausibility of their proposal that consciousness is drenched in self-deception. Fortunately, we do not have far to look for this kind of research today. There is a small but nonetheless suggestive scientific literature on self-deception.

Wishful thinking, the tendency to believe something simply because you would like it to be true, is a type of self-deception to which we are all prone. A large survey of American high school seniors revealed that a full 70 percent thought they had higher than average leadership ability, while a mere 2 percent judged themselves to be below average. All of the one million students surveyed thought they had an above average ability to get along with others. Of these, 60 percent put themselves in the top 10 percent, while 25 percent considered themselves to be in the highest 1 percent. These figures cannot simply be attributed to youthful inexperience: a survey of college professors revealed that all but 7 percent believed that they were better than average at their work.[41]

Sexuality is a rich repository of self-deceptive thinking. In an amusing experimental study that is widely cited in the literature, two groups of heterosexual men were shown a variety of erotic films.[42] One group consisted of men who were comfortable in the presence of homosexuals, while the second consisted entirely of homophobic men. Each group viewed a selection of graphic films depicting homosexual, lesbian, and heterosexual erotica. They were also connected to a plethysmograph, a device that measures subtle changes in the circumference of the penis. The plethysmographic readings demonstrated that the lesbian and heterosexual films excited both groups of men, but that only the *homophobic* men were physically aroused by the homosexual films. However, when experimenters asked them, all of the homophobic men flatly denied that the sight of men having sex with one another had stimulated them. Of course, it is possible that they were lying, but it is also possible that they were deceiving themselves about their own sexual responses. If this suggestion seems implausible now, it will seem less so by the time you have finished reading this book.

Most of us tend to adhere to the outdated Cartesian principle that motivation is first-person transparent: that is, that we all know why we do what we do, despite the considerable evidence that this is not always true. Consider the "bystander" phenomenon beloved of social psychologists. On March 13, 1964, a woman named "Kitty" Genovese was brutally attacked and repeatedly stabbed while walking from the parking lot to her New York City apartment building. Her attacker returned three times during the thirty-five minutes between the first assault and the final, deadly stab, and although she screamed, "Oh, my God, he stabbed me. Please help me!" and later cried out "I'm dying!" not one of the thirty-eight people who observed the scene from their apartment windows bothered to call the police. Later, each of the witnesses said that they had assumed someone else had already called 911. In what has become a classic study of helping behavior, two social psychologists, Bibb Latané and John Darley, were inspired by the Genovese murder to investigate what they called the "bystander effect." In one experiment, they engineered situations in which subjects on their own were faced with someone having a (simulated) epileptic seizure, while in other cases this happened in the presence of bystanders. As it turned out, subjects were less and less likely to come to the person's assistance as the number of bystanders increased, an effect attributed to the "diffusion of responsibility," the assumption that someone else will help. Latané and Darley assumed that the decision not to help when there are others around is a perfectly conscious one, and were astonished to discover that their subjects were *completely unaware of the impact of the presence of other people on their behavior.* "We asked this question every way we knew how: subtly, directly, tactfully, bluntly. Always we got the same answer. Subjects persistently claimed that the other people present did not influence their behavior."[43] The decision to refrain from helping was based on unconscious rather than conscious considerations. Although these individuals believed that their behavior had nothing to do with the presence of others around them, they were clearly self-deceived. The social psychological literature is full of similar examples.

Many mental health professionals promote the idea that depression and other emotional disorders stem in large measure from irrational thinking. Depressives, they claim, believe false ideas about themselves and others. They are self-deceived and out of touch with reality. Irrational, self-deceptive thinking is alleged to be a factor distinguishing depressed people from "normal" ones, but this psychiatric homily turns out to be badly mistaken.[44] Scientific research leads to the opposite conclusion that depressives seem to have a *better* grasp of reality than the "normal" psychiatrists treating them. Lauren Alloy of Temple University in Philadelphia and Lyn Abramson of the University of Wisconsin designed an experiment in which one of the investigators secretly manipulated the outcome of a series of games. Both depressed and nondepressed subjects took part in these fixed games. Psychologists have long known that "normal" thinking involves an element of grandiosity: we tend to give ourselves credit when events work in our favor, but dish out the blame to others when they pan out to our disadvantage. True to form, the non-depressed subjects overestimated the degree to which they had personally influenced the outcome when the game was rigged so that they did well, and underestimated their own contribution to the outcome when they did poorly. Turning to the depressed subjects, Alloy and Abramson found that depressed individuals assessed both situations far more realistically. The rather startling conclusion is that depressives may suffer from a *deficit* in self-deception. Similar results were obtained by the distinguished behavioral psychologist Peter Lewinsohn, who found that depressed people are often able to judge others' impressions of them more accurately than non-depressed subjects are. In fact, these people's ability to make accurate interpersonal judgements *degenerated* as their depressive symptoms diminished in response to treatment.[45] Others have found that high levels of self-deception are strongly correlated with conventional notions of mental health, and that subjects with so-called mental disorders evidence lower levels of self-deception than "normal" people.[46] This research suggests (although, of course, does not conclusively prove) that "normality"—whatever that word means—may rest on a foundation of self-deception. Remove or undermine the foundation, and depression or other forms of emotional difficulty may emerge. If mental health depends upon a liberal dose of self-deception then perhaps, as the philosopher David Nyberg wryly remarks, "Self knowledge isn't all that it's cracked up to be."[47]

If it is true that we are all natural-born liars, it follows that the scientific investigation of human nature runs against the grain of human nature itself. It is triply paradoxical that although we are the only animal that has evolved a mind with the remarkable power to scientifically analyze its own nature, this same mind has been configured by the forces of natural selection to oppose and dismiss the outcome of this investigation.[48] Perhaps the best place to start is to look at the role of dishonesty in other organisms. In doing this, we can begin to understand just how natural deception is and to get a sense of the strategies that we have inherited from our prehuman ancestors as they slowly trudged down the long road of evolution.

2

Manipulators and
Mind Readers

*But the reverse of truth has a hundred thousand forms, and a field
indefinite, without bound or limit.*

—Montaigne

Lying is a natural phenomenon. The biosphere teems with
mendacity. Deception is so widespread amongst non-human
species, so perfectly normal and expectable, that any attempt to
catalogue it comprehensively would be futile.

We should not expect anything different when we look at *human* nature. Deceit is part of our nature, just as it is part of
theirs. If anything, our big brains, intense sociality, and behavioral flexibility imply that we should be capable of deceptive
feats that are far more intricate and devious than anything that
can be observed in the repertoire of non-human species. With
this lineage behind us, it is hardly surprising that human society
is in large measure a densely woven fabric of trickery and dissimulation.

This chapter has two aims (in addition to entertaining you
with some wondrous stories about animal behavior): to prove
that deceit is natural and that we are not alone in our disposition to dishonesty; and to show that even very simple organisms are capable of immensely subtle forms of deception and

manipulation. If even a brainless orchid can wrap a wasp around its metaphorical little finger, then this lends credence to the view that the human capacity for deception, and the related phenomena of deception detection and unconscious communication, must be formidable indeed.

From the lowly virus to the great apes, living things exchange meaningful signals. They engage in conversations expressed in languages of form and color, chemicals, behavior, and sound rather than words and sentences. The moment that you step outside your front door, you are bombarded by a cacophony of messages expressed in a million languages that you do not understand and of which you may not even be aware. The color of the autumn leaves, the song of a mockingbird, the heady cocktail of flower scents, the chemical trail crossing the sidewalk along which black ants busily march, all of these and more are part of a massive conversation.

Once upon a time, biologists believed that the function of animal and plant communication is to transmit true information from sender to receiver. This reassuring, commonsensical idea is wrong. It reflects an idealization of nature, a failure to observe closely and to abandon deeply ingrained prejudices about the moral order of the world. Of course, as soon as we admit that deception is natural, we are just a short step away from recognizing that lying comes naturally to our own species as well. For reasons that will become apparent in future chapters, this is generally very difficult for us to avow.

To flourish, living things must be able to make good use of the resources around them, and these resources include other organisms. If a creature cannot get what it wants from others by exercising force, it must do so by using guile. Some animals apply the so-called puppet master strategy of tinkering directly with another creature's brain. For example, *Euhaplorchis californiensis*, a parasitic fluke that infects banded killifish that live in the coastal marshes of California, causes its host to lose its normal sensitivity to danger. After entering through the gills, this tiny hijacker follows a nerve to the little fish's brain, where it sets up a laboratory to manufacture chemicals that play havoc with the fish's nervous system. Before long, the infected fish starts to behave strangely, spending a lot of time shimmying ostentatiously, swimming near the surface on one side, and exposing its shiny belly to the glint of sunshine. These recklessly suicidal behaviors advertise the fish's presence to shore birds on the prowl for lunch. Infected killifish are consequently four times more likely than their uninfected peers to be snapped up by hungry birds. This is all part of the fluke's devious plan. It tampers with the neural mechanisms responsible for the killifish's swimming behavior because it needs to get into a shore bird's stomach to complete the next stage of its life cycle. A different species, *Crassiphiala bulgoglossa*, accomplishes the same goal using only slightly different means. It performs neurosurgery on the hapless fish to make it antisocial. The infected killifish do not shoal with others of their kind in the presence of a predatory bird, which makes them much more likely to be eaten than would otherwise be the case.[1]

For a really intricate example of direct physiological manipulation, it's hard to beat the rat tapeworm. These repulsive creatures live inside rats, where they grow to a remarkable size, and pass their eggs out with the rats' feces. The rat droppings, which are seeded with tapeworm eggs, are also laced with chemicals that give them an aroma that scavenging beetles find utterly irresistible. Once a beetle consumes egg-laden rat dung, the infant parasites emerge and bore their way into its bloodstream. The immature tapeworms next get to work producing chemicals to meddle with their host's metabolism, preventing it from using its body fat to support the development of its own ovaries, and diverting the fat to feed the ravenous worms. Having waxed strong on the beetle's reserves and sterilized the poor creature in the process, the tapeworms need to make an exit. They arrange this by the simple expedient of manufacturing opiates that make the bug high. The euphoric beetles stumble around slowly, lackadaisically, and oblivious to peril. At this point, only one hurdle stands between the tapeworms and their objective of getting into a rat's digestive system. The scavenger beetles are equipped with a pair of glands that secrete a nasty substance into the mouth of any rat that tries to munch on them. So, in order to maximize the chances of reaching their destination, another rat, where the whole cycle will begin again, the relentless tapeworms chemically disable the beetle's final defense against predation: they knock out its defensive glands.[2]

This kind of invasive physiological intervention is, in the grand scheme of things, a relatively uncommon approach to behavioral manipulation. It is more usual for living things to exploit one another by sending dishonest signals. Manipulating a creature by signals is a very different matter from manipulating it by force. In order to manipulate another by force, an animal need only be strong or fast enough to get what it wants. The more subtle techniques of manipulating others entirely by the deceptive use of signals requires it to find a way to get the animal to move itself in the desired direction.[3] Just as an intuitive "understanding" of the laws of physics allows a chimpanzee to crack nuts with a stone, so an "understanding" of the laws of behavior makes it possible for one organism to convince another to do its bidding.

One option is to use the art of seduction. The pollination strategies of the bee orchids of southern Europe and North Africa provide some amazing examples. One of these, the mirror orchid (*Orphys speculum*) produces small flowers with no nectar to attract potential pollinators. But the orchids have a special ruse to seduce the unwary: they impersonate female wasps of the species that pollinates them.

The blue-violet center of the flower resembles the reflections from the halfway-crossed wings of a resting female. A thick set of long, red hairs imitates the hairs found on the insect's abdomen. The antennae of the female wasp are beautifully reproduced by the upper petals of the orchid, which are dark and threadlike.[4]

In effect, the orchid produces insect pornography to capitalize on the wasp's sexual urges.[5] As is the case in any scam, timing is crucial. Males of the pollinating species mature about a month earlier than the females, and during this period, the orchids are the only action around. The flower's scheme begins with the release of a strong scent that simulates the pheromones (sexually arousing chemicals) released by female wasps. The artificial wasp scent is actually hyper-potent: it is so effective that males are more attracted to the scent of the orchid than they are to that of real females. The intoxicating fragrance, combined with the beguiling shape of the flower, lures the male wasp into an arousing yet ultimately frustrating cross-species erotic entanglement.

In a real copulation, males use the hairs on the female's abdomen to get themselves into the right position for mating. The specific stimuli that would bring about ejaculation are not present, so the wasp, unable to obtain satisfaction, lingers on the flower, picking up more and more pollen in his desperate attempt to breed. Biting the flower in frustration, he only succeeds in releasing more imitation wasp-pheromones to bewitch his senses; and, when he finally leaves, it is not long before he falls under the seductive spell of another flower, depositing on it some of the pollen that is still clinging to his body, and picking up some more, thus unwittingly serving the orchid's reproductive interests.

Mind Reading

The mirror orchid exploits the wasp in a manner that is similar to the techniques used by human scam artists, and just like a human con man, the orchid has to be a good psychologist in order to pull it off. It has to "understand" what motivates its victim, and "use" this knowledge for its own ends. In the lingo introduced by zoologists John Krebs and Richard Dawkins, the orchid has to be a good "mind reader."[6] It may sound strange to speak of a brainless plant being a mind reader and possessing knowledge, but this may be nothing more than a reflection of our parochial human-centered prejudices. Human knowledge may be just a special case of the broader category of biological knowledge, the wisdom expressed in the innumerable ways that organisms are adapted to their environments. "All adaptations," writes University of London psychologist Henry Plotkin, "are knowledge." Adaptations such as the mirror orchid's uncanny ability to manipulate its pollinator express the accumulated wisdom of millions of years of trial and error experimentation, millions of years of evolution.[7]

Life is full of tough decisions. Animals are constantly confronted with choices about how to avoid being someone's dinner, with whom to mate, with whom to fight, from whom to flee, and so on. What makes a choice the "right" one depends on the animal's assessment of the likely outcome. "For an animal that has any kind of social life or who is a predator or who is preyed upon," write Krebs and Dawkins, "these probable consequences will depend crucially on the internal motivational state and probable future behavior of other animals—rivals, mates, parents, offspring, prey, predators, parasites, hosts."[8] In other words, animals must be able to *predict* the behavior of others.

I recently witnessed an impressive demonstration of mind reading when I discovered a family of squirrels nesting in my attic. When I clambered up to investigate the source of the scurrying sounds that were keeping my wife and me awake, the alarmed mother squirrel dived out of a roof vent, but her three adolescent offspring hunkered down. As I scrambled precariously over the beams trying to catch these youngsters, while trying to avoid putting my feet through the bedroom ceiling, I became painfully aware that evading capture was child's play for them. The rodents sized me up, anticipated what I was about to do, and chose evasive tactics that frustrated my every attempt at capture. Only an animal-friendly squirrel trap secured my ultimate triumph.

Now, how did these little creatures consistently manage to outwit a big-brained primate like me? Given the fact that they had no prior contact with predators (they were born and raised in my attic), how could they possibly have acquired these skills? The answer lies in the history of their species. The squirrels' ability to predict my actions was rooted in cognitive adaptations handed down from ancestral squirrels' dealings with predators. They could intuitively "read" me because those of their predecessors who were unable to do so left no descendants to creep into lofts. It is only because ancestral squirrels never had to grapple with wire-mesh traps from the hardware store that their descendants are vulnerable to this deceptive predatory tactic.

Mind reading facilitates deception, and deception encourages mind reading. If one organism knows what another is after, if it is able to divine its mood and second-guess its reactions, this opens the door for exploitation and manipulation. Similarly, an ability to read minds gives protection against manipulation. Mind reading and manipulation bounce off each other in an intimate dialectic spiraling through evolutionary time. You will discover more about this relationship in chapter 3.

Poker-Faced Liars

One way to fool others is to become invisible. Camouflage, or crypsis as biologists call it, is the art of concealment. It enables an animal to hide in plain sight either by blending in with its background or by disguising itself as some particular object that is of no interest to the creature from which it is hiding. One common technique is changing color to match the background, associated in the popular imagination with a surreal little African lizard called the chameleon. In fact, it is also used by worms, mollusks, insects, spiders, fish, amphibians, and even birds and mammals. Color shifting is not the only kind of quick-change artistry. Some species of cuttlefish, for example, are able to modify the texture of their body surface and can even sprout new appendages to become indistinguishable from their environment.

In mystery films, the victim often notices a concealed murderer by the shadow he casts. This dramatic device is biologically realistic: shadows can be a dead giveaway to predators and prey alike. Some animals have dealt with this problem by evolving flattened body forms that don't cast much of a shadow. Others, such as the domestic cat, flatten their body as much as possible when advancing on their prey. Countershading provides yet another wrinkle on color camouflage. A countershaded animal is darker on its back than on its belly. This arrangement makes sense if you consider the fact that light almost always strikes an animal from above, so that its back is usually more highly illuminated than its underside. Having a back that is *darker* than its underside compensates for this factor and helps an animal to blend in with its environment more effectively. A third technique, disruptive coloration, exemplified by the wild pattern of a zebra's stripes, is very useful for herd animals because then it is difficult for a predator to tell where one animal ends and another begins.[9]

A problem with most forms of camouflage is that they are only effective when the animal are standing still. No matter how perfect the disguise, as soon as the animal takes a step, the spell is broken and it stands out like a sore thumb. This presents a special problem for predators that need to stalk their prey. The usual solution is to creep up slowly, so that your victim does not notice that you are moving until the final sprint at close range. The zone-tailed hawk (*Buteo albonotatus*) has hit on an alternative approach: it uses a form of *moving* camouflage. The hawk blends in with a crowd of turkey vultures languidly cruising along in search of carcasses. Being scavengers, vultures are no threat to the small birds and mammals upon which the hawk feeds. It is only after the vultures happen to glide near to the prospective prey that the hawk breaks cover and swoops in for the kill.[10]

The forms of specific camouflage are extraordinarily varied. Creatures specialize in resembling twigs, leaves, flowers, coral, pebbles, and even splattered bird droppings. However, crypsis is a relatively simple form of deceit that shades over into something much more complex: the art of mimicry.

Mimics, Models, and Dupes

An alternative to invisibility is to assume a disguise, take on a false identity, and become a mimic. Mimicry is a con game involving three parties: the mimic (who performs the deception), the model (whom the mimic pretends to be), and the dupe (who is hoodwinked). Biologists have identified several forms of mimicry, although this categorization is probably not exhaustive.[11]

The history of the study of mimicry begins with the Victorian naturalist Henry W. Bates. While collecting specimens deep in the Amazonian rain forest, Bates noticed a strong resemblance between two entirely unrelated kinds of butterflies. Heliconid butterflies have an awful taste that birds avoid, but the strikingly similar Pierids are an avian delicacy. Bates inferred that the Pierid butterflies had evolved to resemble the Heliconids because the similarity protected them from birds. Batesian mimicry, as it came to be called, occurs when an innocuous creature imitates a nasty one.

The little bushveld lizard (*Heliobolus lugubris*) is native to the Kalahari Desert of southern Africa. Adults are buff-colored and camouflage well against the desert surface. The young, however, are black to mimic the thoroughly unpleasant oogpister beetle, which deploys an arsenal of pungent acids, aldehydes, and other chemical weapons against predators. By impersonating oogpisters, the vulnerable little lizards benefit from the beetle's deservedly bad reputation. The impersonation shows careful attention to detail. Not only are the immature lizards the same color and size as their model, but also they imitate the beetle's jerky, stiff-legged robotic walk. It is only after they grow to the size of the largest oogpisters, beyond which point the fraud will no longer work, that the youngsters abandon their beetle mannerisms and metamorphose into the adult color.[12]

The show put on by the North American hognose snake (*Heterodon platirhinos*) is one of the world's most extravagant examples of Batesian mimicry. When approached by an erstwhile predator, such as a human being, this nonpoisonous snake pretends to be angry and dangerous to be near: it flattens its head, spreads out a hood resembling that of a cobra, and hisses violently. Sometimes it pretends to strike, with maniacal aggression, all the while keeping its mouth discreetly closed. If this tactic falls short of terrifying the interloper, the snake performs a melodramatic death scene, writhing in agony and finally flipping over on its back and "expiring" with mouth gaping and tongue lolling out pathetically. For added pathos it may even drip blood from its eyes and emit a disgusting smell apparently to give the impression that it is not merely dead, but is actually rotting! If one tries to spoil the effect by reaching down and turning the snake over, it will immediately flip back into the belly-up "dead" position.[13]

The greatest Batesian impersonator is a two-foot-long octopus native to the waters off Sulawesi and Bali, Indonesia. The mimic octopus, which at the time of writing has not yet been assigned a scientific name, is a true master of deception, with an astounding repertoire of impersonations of toxic species. Sometimes it disguises itself as the venomous sole, a flat fish native to these waters, by shooting through the water with its arms arranged in a disk-like configuration and its body undulating in a fish-like swimming motion. At other times, the octopus dons the garb of the spiky, noxious lionfish by spreading its arms in a wonderful imitation of the lionfish's fins and floating along just above the bottom. In the blink of an eye, the octopus disguises itself as two sea snakes by burying six of its tentacles in the mud and leaving two free to wave in the current. The tentacles change color, developing the black and white bands characteristic of their model, the snake. This versatile deceiver may also be able to impersonate various other sea creatures.[14]

Aggressive or "Peckhamian" mimicry uses the Little Red Riding Hood method: a predator disguises itself as something attractive to its prey. The alligator snapping turtle (*Macrochelys temmincki*) is a large and extremely menacing looking reptile native to the American South. Snappers are massive reptiles, weighing up to 250 pounds. Obviously, they are not built for high-speed chases, nor do they have to be. The snapper spends its days lying inertly at the bottom of a creek or bayou, where it resembles a rotten log, with its large mouth wide open. The lining of the turtle's mouth is gray, but right at the back there is a vibrant pink imitation worm that beckons invitingly to passing fish. Any fish foolish enough to investigate is either swallowed whole or sliced in half by the turtle's razor-sharp jaws. Some varieties of the extraordinarily bizarre frogfish (family *Antennariidae*) that live in the perpetual twilight of deep water are equipped with a long movable spine attached to the head that acts as a fishing rod festooned with glow-in-the-dark artificial bait. Just like an expert angler, the deep-sea frogfish moves the bioluminescent lure in a pattern calculated to interest its prey, while releasing chemical attractants into the water. When some poor creature approaches for a nibble, the frogfish opens its vast mouth and with unbelievable rapidity (one six thousandth of a second) devours its prey in one magnificent gulp.[15]

Some fireflies of the genus *Photuris* use a femme fatale variant to lead males of other species to their demise. Fireflies use their luminous abdomens to flirt. Flashing out a sexy message that they know will titillate males of the target species, these deadly ladies attack their suitors, feast on their flesh, and rob them of defensive chemicals.[16] Bolas spiders (*Mastophora hutchinsoni*) churn out imitation moth pheromones to attract a meal. These small creatures spin a single strand of silk, a fishing line at the end of which is a sticky "hook" soaked with simulated moth pheromones. When the spider feels the wing beats of an approaching moth, it waves its silk trap in the air rather like an angler casting his line. Once snared, the spider has only to reel in the moth.[17]

Jumping spiders of the genus *Portia* are true virtuosos of deception. In the first place, they don't look like spiders at all: they resemble pieces of debris, which is a great way to deceive both predators and prey. Unlike most spiders, which feed on insects, the button-sized portia spider preys on spiders up to twice its size. To understand its refined hunting methods, it is important to realize that most spiders have lousy eyesight but are exquisitely sensitive to the vibrations of their web and interpret specific patterns well enough to distinguish between, say, the quivering produced by an ensnared grasshopper struggling to free itself and the movements of a dead leaf blown onto the web's sticky surface. Portia spiders have unusually superb eyesight and, as we will see, use their eight eyes very effectively when stalking their kill.

Once the spider zeros in on a spider that it fancies eating, it gingerly creeps into the victim's web and adroitly plucks at the threads to simulate the effects produced by a trapped insect. Now, each type of spider has a different way of interpreting the vibrations in its web, so in order for this sleight of hand to be successful, the portia spider must identify the species it is hunting and use the correct species-specific pattern. An instinctive knowledge of many of these patterns is hard-wired into the portia's minute brain, so once it makes a correct identification, it can adjust its efforts accordingly. What if the spider finds itself stalking a species for which it lacks a preset program? No problem. It tries out sequences of random variations, while keeping a close watch on its quarry; and, as soon as the latter responds by edging closer, our protagonist stops generating random variations and concentrates on repeating the one pattern that produced the desired effect. This is quite a delicate operation. If the intended victim is too large and strong, imitating its prey could result in a rushed attack, turning the tables on the would-be predator. The spider preempts this outcome by carefully and systematically controlling its mark's behavior. It uses its sharp eyesight to estimate the size of its prospective victim, observes its movements, and uses this information to fine-tune its tactics in a game of cat and mouse that can last for hours. Dangerous prey are drawn in slowly, calmed by hypnotically monotonous web vibrations, and steered into an orientation which gives the predator the greatest attack advantage. The spider may even make elaborate strategic detours, including those that require it to move away from and temporarily lose sight of its prey. Sometimes the portia spider climbs onto an overhanging branch and lowers itself on a single silk thread to within striking distance of its unsuspecting quarry. The other spider, feeling no vibrations in its web, is oblivious to the presence of a deadly assassin.

It is particularly advantageous for the portia spider to approach in a roundabout fashion when dealing with spitting spiders, which are able to spew their venom at objects up to ten body-lengths in front of them. In the Philippines, where portia spiders prey on spitting spiders, they make extensive use of the detour method to approach dangerous prey from behind. Female spitting spiders carry their egg-sacks around in their mouths which prevents them from spraying poison at would-be predators. Portia takes account of this chink in these spiders' armor, and attacks egg-carrying females from the front. When the wind blows or raindrops, twigs, or leaves fall on the web, the predator cleverly uses the resulting vibrations as a smokescreen for its own movements and advances quickly toward its prey. In less hospitable circumstances, the portia spider generates its own "noise" by shaking the web in patterns that mask the telltale effects of its own approach. Not bad for an animal with a brain considerably smaller than the head of a pin![18]

The strategy known as social parasitism sometimes requires the perpetrator to be an expert impersonator. Consider the little staphylinid beetle (*Atemeles pubicollis*), who manufactures chemicals to dupe the ants off whom it sponges. Ants are ordinarily able to discriminate between their own kind and strangers by detecting species-specific chemicals called allomones. Ants are often highly xenophobic and deal very harshly with unauthorized visitors. To get past the border patrol, the beetle mixes up a little potion of allomones—in effect forges a chemical passport—and sprays it at a guard. This induces a delusion that the beetle is actually an ant larva (a baby ant), and the ant gently picks the intruder up and deposits it in the colony nursery. Once inside, the beetle lives the life of Riley: the ants cater to its every whim, feeding, pampering, and protecting it. The beetle has the run of the colony, where it mooches food from the ants as well as snacking on ant eggs and larvae with impunity.[19]

Many animals, including human beings, are dedicated followers of fashion who like to mimic other members of their own kind. Biologists call this behavior automimicry. There are three forms of automimicry: mimicking one's own body parts, mimicking one's entire body, or mimicking other members of one's species. There are creatures like the African two-headed snake, whose rear end looks just like a second head. Two heads are better than one because an attack to the fake head is less likely to be lethal than an attack to the real head. Many lizards have detachable tails, which come loose when grasped or are even snapped off at will. Once parted from the lizard's body, the tail writhes frantically on the ground as a decoy, impersonating the whole lizard.[20] Sepiolid squid also manufacture replicas of their whole body. When the squid detects a potential predator, its first move is to change its appearance by darkening its skin. Although this may seem counterproductive because it makes the creature stand out more from the sandy ocean floor, there is method in the mollusk's madness. The squid next squirts a billow of ink out of its rectum. The ink is composed of melanin, which gives it a dark color, and mucus, which prevents it from diffusing freely through the water. (It also contains tyrosinase, an irritant that temporarily disables the predator's sense of smell.) The squid releases just the right amount of ink to create a cloud roughly the same size and shape as itself, and simultaneously changes to a lighter color and slips unobtrusively away. With this "pseudomorph," the clever mollusk has effectively reduced its chances of being caught and has left the unlucky predator with only a puff of ink to eat. Because this inky approach to deception would be useless in the dingy conditions of deep water, sepiolids that inhabit the ocean depths discharge a *luminescent* cloud to distract predators.[21]

Automimicry can also be handy for attracting a mate by presenting oneself as a more desirable specimen. Male three-spined sticklebacks (*Gasterosteus aculeatus*) show their machismo by exhibiting a red patch on their sides. The fish's display is meaningful because the red pigment comes from beta-carotenoids, a valuable nutrient in these fish's diet. Throwing away beta-carotenoids for the purpose of display is conspicuous consumption, like lighting a cigar with a fifty-dollar bill: it is a sign of having ample resources aimed at prospective mates who want their spawn to inherit high-quality genes. Female sticklebacks find it irresistible. The females are quite right to do so: a bright red patch advertises dominance, health, and the ability to protect a clutch of eggs from other fish. But these signals are not always what they seem. Some lower-quality males display red colors to mislead females. These undernourished Romeos are often in such poor shape that if they manage to trick a female into mating with them, they often cannot resist gobbling up their own clutch of eggs instead of guarding them against predators. When one dominant stickleback spots another, its nuptial patch acts like the proverbial red cape to a bull. So, when confronted with the genuine article, the impostors make their spots fade to avoid a nasty confrontation with the more virile fish.[22]

Cross-dressing is another form of automimicry found in a variety of species (including, of course, our own). The giant cuttlefish (*Sepia apama*) is a heavy-bodied squid-like creature that lives in the shallow coastal waters off the coast of southern Australia. During the summer, these animals congregate in large groups to breed. Competition between the muscular yard-long males is intense and often violent. During these sex tournaments, small males loiter in the vicinity of the breeding pairs. These weaklings are not attacked by the big bruisers because they are not even recognized as males: the small males mimic the color and pattern of females, and even hide the fringes around their tentacles (a cuttlefish gender giveaway). It turns out that when Tarzan is off beating up another cuttlefish, the little guy reverts to his male colors and has sex with Jane. Should the dominant male return unexpectedly and catch them in flagrante delicto, the smaller animal immediately switches back to his female persona.[23]

A species of the parasitic wasp (*Cotesia rubecula*) has an unusual variation on this theme. Sexual competition among male wasps is intense. This is aggravated by the fact that female wasps remain in the mood for love for a short period after a prior mating. Thus, a wasp who has recently mated may well end up being reproductively trumped by a competitor. To avoid this, a male wasp that has just copulated may mimic an attractive female in order to distract the other males from his mate until her sexual receptivity wanes.[24]

Another good reason to impersonate a female is to get rich quick. Among scorpion flies of the species *Hylobittacus apicalis*, it is de rigeur for a male to present a female with a gourmet meal consisting of a fresh insect corpse before she will even consider mating with him. Because the female is busy enjoying her meal while he is having intercourse with her, the size of the offering is important. A good-sized feast can lead to twenty minutes or more of lovemaking, whereas something more meager is good only for a quickie, if not complete rejection. Hunting for nuptial bribes is hard and dangerous work, and some males opt for a shortcut. They disguise themselves as females, and when an unsuspecting male sidles up to "her" with a juicy morsel, the male-in-female's clothing snatches it and flies off to offer it to a mate of his own.[25]

In a particularly mind-boggling form of automimicry, flying dragonflies imitate themselves standing still. Picture an idyllic summer scene, with dragonflies flitting above the calm surface of a sun-drenched pond. This placid setting conceals a violent reality. During the breeding season, male dragonflies are fiercely territorial. All of that delicate hovering and darting is actually the behavior of males locked in savage aerial dogfights over prime real estate. These insects use equipment of such sophistication that it puts our own stealth aircraft to shame. Dragonflies judge whether or not an object is moving based on what is called optic flow, the movement of an image across their retina. Akiko Mizutani of the Centre for Visual Science at the Australian National University in Canberra has demonstrated that a dragonfly in hot pursuit of a rival shadows it in the air with such accuracy that the aggressor appears to be stationary in his rival's field of vision. To do this, the attacking dragonfly has to use ultra-precise flight adjustment and positioning to maintain the same position in its victim's visual field from one split second to the next. Male hoverflies use the same tactic when tracking mates.[26]

Finally, certain species use cross-dressing to keep them warm. Biologists studying red-sided garter snake (*Thamnophis sirtalis parietalis*) populations in Manitoba, Canada, found that some males groggily emerging from eight long months of hibernation produce a female pheromone to attract the attention of other males. The she-males quickly become the center of huge mating balls, entwined by dozens of amorous males. Serpentine drag queens are not after sex: they are after heat. The garter snake breeding season comes immediately after hibernation has made them weak, cold, and vulnerable to predation. What better way for these snakes to thaw out than to have one hundred passionate male bodies wrapped around them? Being at the center of a serpentine sex scrum is enough to raise their body temperature by as much as three degrees centigrade.[27]

Although less common, there are examples of female-to-male cross-dressing in non-human species. In some insect species, this appears to be a way of avoiding sexual harassment by overly enthusiastic males.[28]

Lie Detection and Confusing the Enemy

Being a good mind reader is the best protection against manipulation. A species that develops abilities to penetrate a deceptive front renders the deception obsolete. The evolution of increasingly sophisticated mind-reading abilities as a defense against deception has been particularly important for the evolution of mind, something we will discuss more fully in chapter 3.

Mind reading is not the only countermeasure against hostile manipulation. Espionage provides an organism with inside information about a competitor. Various birds and fish of both sexes size up the prowess of their neighbors by snooping on them.[29] Another way to counteract manipulation is to throw sand (metaphorically, and, in the case of some octopi, literally) in a predator's eyes. We have already encountered an example of this in the behavior of the versatile portia spider, which generates its own "noise" to serve as a smokescreen for its movements across its prey's web. The tiger moth provides another fine example. Bats prey upon tiger moths, which they hunt using their well-developed sonar system. As the bat flies along, it emits high-frequency sounds and listens for the returning echoes, which it interprets to determine the size and trajectory of objects in its flight path. The tiger moth's body is studded with "ears" that are tuned to just the right frequency range to hear bat sonar signals from as far as forty meters away. As soon as the moth detects a bat headed its way, it flies in the opposite direction. As the advancing bat goes into a "feeding buzz" sending out rapid-fire echolocation signals to home in on the precise position of the moth, the moth takes evasive action by flying in wild, spiraling loops in an attempt to outmaneuver the clumsier bat. If the bat's sonar remains locked on its target, the moth's last resort, with only milliseconds between it and certain death, is to jam the bat's sonar by emitting a series of clicks that the bat cannot distinguish from its own echoes. In creating these phantom echoes, the moth is often able to confuse the bat and get away. Several species of bats have developed methods of outsmarting the moths' detection and jamming capabilities. These include huge ears that are able to detect the presence of moths and other insects on the ground, soft, "whispering" sonar that the moths can't hear, or signals in frequencies too high or too low for the moths to detect.[30]

A final way of throwing a predator into a tizzy is to behave in a bizarre, erratic, or seemingly random manner. The aerial pyrotechnics displayed by moths, butterflies, and other flying insects to evade birds and bats; the zigzag trajectory of a rabbit chased by a fox; the unpredictable back-flips made by brown shrimp to escape marauding cod; and the weird convulsive movements of frightened laboratory rats and salamanders are all examples of the "protean" defense. This strategy, memorably portrayed in Luke Rhinehart's fictional *Dice Man,* was used by submarine commanders during World War II who tried to confuse the enemy by sailing a random course determined by throwing dice.[31]

A Hierarchy of Deceptions

It may be useful at this point to consider the relationship between the deceptive features of nonhuman species and the forms of dishonesty favored by our own. There are obviously some striking resonances, but there are also some clear distinctions: a human gender bender is doing something rather different than, say, a cross-dressing garter snake, even though both are instances of automimicry.

Some organisms use a deceptive physique. Most forms of camouflage fall into this category, as does a good deal of mimicry (for example, the mirror orchid). Other modes of deceit are switched "on" and "off" by environmental variables. Whereas the counter-shading of a gray squirrel stays the same no matter what the squirrel is doing, the tiger moth's sonar jamming mechanisms only swing into action when a hungry bat is closing in. In the case of the portia spider, we find an immensely more flexible deceptive repertoire. The spider's deceptive behavior is not an "all or nothing" response: it varies systematically with the specific circumstances in which the spider finds itself. Although the portia spider is a very clever and flexible deceiver, it can only use this ability to hunt prey. Other creatures, notably human beings and some non-human primates, can deceive in a variety of situations. We *Homo sapiens* are able to lie across the board. Furthermore, we use these gifts to manipulate our own kind; enemies and friends, lovers and rivals, parents and children. As we will shortly see, this powerful and dangerous combination is one of the factors that gives human social life its distinctive stamp, and has been the driving force behind the evolution of both self-deception and the unconscious mind.[32]

3

The Evolution of Machiavelli

We are primates who are experts in deceit, double-dealing, lying, cheating, conniving and concealing.

—Sanjida O'Connell

At the dawn of the nineteenth century, an Anglican pastor named William Paley described in his influential book *Natural Theology* how living creatures fit their environments with breathtaking precision. For Paley, as for many thinkers before and after him, this was proof that a master Designer had crafted the natural world. In a famous passage, Paley invites the reader to accompany him on an imaginary country walk. "Suppose," he conjectures, "I pitched my foot against a stone, and were asked how the stone came to be there: I might possibly answer, that, for anything I knew to the contrary, it had lain there forever. . . ." He beckons us to contrast this with the experience of noticing a pocket-watch lying on the ground. How might the watch have come to be there?

I should hardly think of the answer which I had before given,—that, for any thing I knew, the watch might have always been there. Yet why should not this answer serve for the watch as well as for the stone? Why is it not as admissible in the second case, as in the first? For this reason, and for no other, viz. that when we come to inspect the watch, we perceive . . . that its several parts are framed and put together for a purpose. . . . This mechanism being observed . . . the inference we think is inevitable, that the watch must have had a maker: that there must have existed, at some time, and at some place or other, an artificer who formed it for the purpose which we find it actually to answer: who comprehended its construction, and designed its use.[1]

The exquisitely adaptive features of living things make them more like watches than like stones. Paley used biological evidence to defend the traditional idea that organisms were fabricated by an omnipotent God and that their adaptive design is evidence of His handiwork. This line of reasoning was a nineteenth-century rendition of an ancient theory of the origins of the natural world. Creationism, the idea that the world came into being by divine fiat, is probably as old as human thought and was a rather good theory for a prescientific age. Living things do give every appearance of having been intelligently designed, and it is clear that no mere human being could possibly have done the job. Given this, and having no reasonable alternative explanation, it was logical to conclude that the author of the biological world was an inflated version of our own species, a God fashioned in our own image.[2]

At the same time that the Reverend Paley was penning the manuscript of *Natural Theology,* other thinkers were beginning to advance purely scientific explanations for biodiversity and adaptation. The idea of evolution, as it eventually took shape in the late eighteenth and early nineteenth century, held that species are mutable rather than fixed for all time, and that they are gradually shaped and reshaped by the impact of a changing environment. The most prominent of the early evolutionary theorists was Jean Baptiste Pierre Antoine de Monet de Lamarck, professor of invertebrates at the Muséum National d'Histoire Naturelle in Paris. (He invented the term "invertebrates" so he wouldn't have the unattractive title of Professor of Worms and Insects.) Lamarck proposed that environmental pressures operating over very long stretches of time transform living things so extensively that entirely new species emerge. The rather comical paradigm of Lamarckian evolution is his reconstruction of how the giraffe acquired its long neck. During a bygone era, a shortage of grass forced the short-necked ancestors of the modern giraffe to dine on leaves, and this required them to stretch their necks. All of this neck-stretching exercise in turn caused their offspring to be born with slightly elongated necks. The long neck of the modern giraffe was supposed to be the upshot of many, many cycles of neck stretching, neck growth, and reproduction. As crazy as Lamarck's theory sounds to contemporary ears, it gave a non-miraculous explanation of the origin of species that accounted for why organisms fit the environments in which they live. Unfortunately, the theory also contained a fatal flaw: individuals do not inherit traits that their parents have acquired through learning or practice. Learning is not transmitted through our genes.

Charles Darwin discovered the real process driving evolution. His breakthrough was the stunningly simple and yet immensely powerful formula of evolution through natural selection. Before Darwin, it was widely believed that it was only a few thousand years ago that God had created all of the species of plants and animals, and that species were immutable and fixed for all time. Humans beings, too, were thought to have been specially created and occupied a unique position apart from (and, of course, above) all other species. Darwin's demonstration that all species, including *Homo sapiens*, evolved from one another over immense expanses of time decimated this comforting vision of the natural order.

Darwinian theory is simple and elegant. It rests on three pillars: variation, selection, and reproduction. Individual variation is the norm in nature. With the exception of monozygotic (identical) twins, no two members of any species are biologically identical. Some of these variations are detrimental; that is, they decrease the likelihood that the individuals bearing them will flourish and reproduce. Other variations are irrelevant to survival prospects and reproductive success (for example, the particular shape of one's eyebrows). Still others are beneficial. Every population contains individuals endowed with physical or mental gifts that give them an edge over their rivals. These stronger, smarter, faster, or sexier individuals are able to produce more surviving offspring than the others and these offspring are likely to inherit that *je ne sais quoi* responsible for their parents' success. In Darwinian parlance, they are more "fit" than others.

Natural selection is a consequence of variations in fitness. It is easy to see how, given enough time, heritable fitness-enhancing traits will radiate through a population, squeezing out the competing alternatives. Characteristics are "selected" to survive in a vast, gruesome version of the game of musical chairs. Preferentially replicated successful traits, selected by nature to survive—at least for the present—are "adaptive" traits. In biological jargon, something is adaptive if it *works* as a solution to a problem in living.

Imagine that a cross section of the population consisting of, say, one hundred individuals, is transported to a remote, uninhabited tropical island that I will call "Darwinia," and its residents, "Darwinians." Each member of the party is a variation on the human theme: tall or short, athletic or sedentary, beautiful or ugly, black or white. Now, imagine that Darwinia is also home to a population of large predators that soon discover that they relish the taste of human flesh. All things being equal, the more athletic members of the human population, those who are most effectively able to run away from danger, will be more likely to survive than the slower, more inert ones. Because (unlike certain parasites) it's not possible for members of our species to have sex inside a predator's belly, the faster Darwinians will be more likely to live long enough to reproduce than their slower peers. Consequently, they are more likely to pass on to their children the genes responsible for their athletic prowess. Under these circumstances, speed is adaptive (because it solves the problem of how to avoid being devoured) and it is fitness enhancing (because it contributes to its bearers' reproductive success). As long as the selection pressure remains in place, the same filtering process will happen repeatedly with each successive generation as the genes for speed proliferate through the population. Eventually, after many stages in successive descent, the average Darwinian will be able to run faster and longer than his or her ancestors.

Nature is wildly prodigal. Most species spawn far more offspring than their environment can support, the vast majority of which are doomed to an early destruction. Consider the Pacific salmon. A single salmon may produce several million offspring in a single breeding season, but only .003 percent of these live long enough to reproduce.[3] In other words, 99.997 percent of salmon never pass on their genes to the next generation. Of course, some of this is just bad luck, but luck is not the whole story. Some manage to survive because of characteristics that stack the deck in their favor, an effect amplified by countless iterations. Perhaps they swim a bit faster or are a tad more resistant to disease; perhaps they prefer slightly colder water that takes them out of range of a predator; or, maybe their coloration makes them less visible. The possibilities are as varied as life itself. What counts is that they possess some heritable feature that makes a difference, and they live long enough to pass it on.

An untimely death is by no means the only reason why a creature may fail to pass on its genes. After all, it is possible to live to a ripe old age and never have sex. In order to be reproductively successful, one has to be attractive enough to instill desire in members of the opposite sex or intimidating enough to repel same-sex competitors. So, nature selects not only for those features that enhance survival, but also for those that are sexy. The "survival of the prettiest" as psychologist Nancy Etcoff puts it, is the force behind the evolution of the extravagant tail feathers of the male peacock and the gentle curves of a woman's body.[4] Darwin called this process "sexual selection."

The massive elephant seal (*Mirounga angustirostris*) dramatically illustrates the power of sexual selection. Elephant seals congregate once a year to mate on the beaches of Mexico and California. Over ninety percent of bull seals live a life of involuntary celibacy; but, the remaining nine percent—the alpha bulls—are fantastically successful, managing harems consisting of up to 150 females. With so many mates, sexually active males must be constantly vigilant to prevent single males, who opportunistically prowl the periphery of the harem, from getting in on the action. In the world of elephant seals size definitely matters, for the greater the bull's bulk, the more effectively he can intimidate competitors. Thanks to sexual selection, elephant seal bulls are enormous hulks—sumo wrestlers of the pinniped world—weighing in at 5,000 pounds and growing up to about sixteen feet in length.

Natural selection holds sway even among species with dramatically lower mortality rates than the Pacific salmon and with more egalitarian sexual arrangements than the elephant seal. This is because any genetically heritable trait that differentially affects reproductive success, however incrementally, will, given sufficient time, inevitably spread. Darwin's greatest achievement was to show how purely random variation could give the misleading impression of intentional design. Natural selection, as Richard Dawkins memorably put it, echoing and subverting Paley's example, is a *blind* watchmaker.[5]

Selfish Genes, Unselfish People?

Although the theory of evolution through natural selection is extremely powerful, it was at first unable to explain *why* organisms vary or give a coherent account of *how* traits are passed on from parents to offspring. Darwin's solution to the problem of inheritance was at once conventional and inadequate. He believed that traits from both parents were "blended" in their offspring. It is not difficult to see that this cannot be right. In the first place, it is inconsistent with observation. When a blue-eyed man and a brown-eyed woman produce a baby, the child does not have bluish-brown eyes. Inheritance seems to operate digitally: its effects are either/or rather than both/and. The blending hypothesis is also blatantly incompatible with Darwin's own theory of evolution through natural selection, because it implies that novel adaptive traits become diluted with each new generation, eventually becoming so watered-down as to be completely blended away. When Darwin's French translator Clémence Royer noted this awkward contradiction, the great man promptly terminated her employment.[6] So much for scientific detachment!

Although he could not have realized it, Darwin needed the science of genetics to understand how inheritance works. While Darwin was refining the theory of evolution in England, a Czech monk named Gregor Mendel was laying the foundations for genetics through his dogged research selectively breeding pea plants in the monastery garden. Mendel was the first person to infer that there must be discrete hereditary elements (now known as genes) passed on from parent to offspring. Mendel's hereditary units were theoretical entities: he knew that they must exist because the precise patterns of heredity that he had observed would be inexplicable without them, but he had no idea what sort of physical structures carried hereditary information. We now know that genes are sequences of molecules strung like Christmas-tree lights along microscopic structures called chromosomes. When sex cells multiply, genetic information is randomly "shuffled" endowing each daughter cell with a unique permutation. Although this information is usually copied accurately, once in about every 100,000 replications there is a transcription error, or mutation. Mutations create novel genetic information. Most of them are damaging (to understand why, just imagine randomly modifying the circuitry of your computer—it is tremendously unlikely that any given modification would actually improve the computer's performance), but occasionally one occurs that is beneficial. Once introduced, mutations replicate like any other piece of biological information.

Mendel's work had little impact during his lifetime. Rediscovered at the turn of the twentieth century, it was not yoked firmly to Darwinian theory until the 1930s in a movement known as the Modern Synthesis. Genetics provided the missing link needed to complete Darwinian theory because it supplied an explanation of the generation, selection, and reproduction of heritable characteristics. Those organisms carrying genes that give them an adaptive advantage over their less fortunate peers are likely to survive long enough, or be sexually attractive enough, to successfully reproduce the advantageous gene, which is then inherited by that creature's offspring. This is how purely random mutations that fortuitously equip an individual with an adaptive benefit manage to proliferate through a population. Ultimately, then, genes are the units chosen by natural selection. Those genes that are best able to secure their own continued reproduction remain. The failures vanish.

Nature selects for reproductive success. Those characteristics that enhance health, survival, and sexual opportunity are most likely to pass from one generation to the next, while features that detract from reproductive success are ruthlessly weeded out. This *seems* to imply the world should be full of self-interested, egotistic creatures and that any inclination to unselfishness should have long ago gone into the genetic garbage can. But this is clearly not the case. Human beings are certainly capable of self-sacrifice and self-denial for the benefit of others. Think of parents who put their own desires on the back burner for the sake of their children's well-being, or fighters for a cause, or humanitarians who help those in need. These well-known examples of human altruism appear to fly in the face of Darwinian theory.

Does the existence of altruism prove that human beings are "higher" than mere animals, that we, unlike them, can transcend the brute laws of evolution? Altruism cannot be used as a brick in the wall setting off *Homo sapiens* from all the rest because many nonhuman creatures display altruistic traits. One such example is babysitting, a relatively common form of altruistic behavior. The dwarf mongoose (*Helogale parvula*), a small, ferret-like carnivore found in Tanzania, is one of many species that looks after infants that are not their own. When mom and dad are out hunting, an older sibling supervises the kids. The caretaker thus puts in time that it could perhaps invest more profitably in other, directly fitness-enhancing activities, and may even have to put its life on the line defending the burrow from predators. Alarm calling is another example of nonhuman altruism. Group-living species have the advantage of many sets of eyes and ears to detect approaching danger. An individual that spots a predator nearby will often sound an alarm call, signaling the others to dive for cover. This behavior runs counter to raw self-interest because the animal that sounds the alarm also attracts the predator's attention. If nature only selects behavior that directly contributes to an individual's reproductive success, such animals would be inclined to slip quietly away instead of noisily making their presence so obvious.[7]

Reproductive altruism is perhaps the most striking form of altruism found in nature: an individual entirely foregoes reproduction, apparently to promote the reproductive success of another. The social insects, most notably the ants, which have been spectacularly successful in the struggle for existence, are the paradigmatic example of kin altruism. At least one in every 1,000 insects is an ant, and in some parts of the world, the biomass of the ant population exceeds the biomass of all vertebrates combined. Ants live in societies with sharp divisions of labor. Reproduction is the unique privilege of the queen and her male harem, while virginal female workers attend to the everyday tasks of the colony. The daily drudgery of the worker ant does nothing at all for the reproductive prospects of the worker herself, and seems only to benefit the queen whom she serves. Many of the workers' activities are highly differentiated and sophisticated. In some species workers practice animal husbandry, maintaining herds of aphids that they "milk" for nutrients. Others have a farmer caste. Using leaves, which they cut and carry to the nest as a growing medium, fungus-growing ants plant fungus, "weed" the fungus patch, fertilize it, and harvest the edible part of the fungus that they have grown. Still other species have workers that specialize in slavery, waging war on neighboring ant colonies and abducting larvae and pupae that grow up to be their slaves.[8] Now, here is the puzzle. In order for a trait to spread through a population and become established in a species, just one thing is required: the trait must improve the chances that the individual possessing it will reproduce the relevant gene. Yet, altruistic behaviors do not load the reproductive dice in favor of the individuals concerned. Worker ants tending their herds of aphids, farming fungus, and enslaving others are *unable to reproduce*. Does this not demonstrate that there is something seriously wrong with evolutionary theory?

Biologists once commonly believed that traits like altruism are selected for the good of an entire community of organisms. This theory, known as group selectionism, claims that natural selection has seen to it that at least some individual interests are subordinated to those of the group. These traits benefit the social unit as a whole, even if they are detrimental to the individuals concerned. A group selectionist might propose that genes responsible for the altruistic behavior of the worker ant were present because they enhance the survival of the entire colony rather than any benefit they provide to the individual ants.

Although this might sound plausible enough, it is probably wrong in most biological contexts. Debates regarding the problem of "levels of selection" tend to be rather arcane, but the gist of the difficulty is called the free-rider problem. Imagine a population living in an environment where the amount of food available can optimally support a group of n individuals. If the population rises above this level, then members of the group begin to go hungry and to die. Imagine, too, that a process of group selection has operated on this population so each individual limits their fecundity to keep the population at this optimum level, even though it would be advantageous for any individual to reproduce without inhibition. A moment's consideration shows that such a population would not have very good survival prospects. Unless *completely* isolated, the group would be in constant danger of infiltration by individuals that do not possess the gene for reproductive restraint. These interlopers would reap significant benefits: while everyone else was conscientiously practicing family planning, the newcomers would be breeding with abandon. With each new generation, the percentage of selfish breeders would swell and the proportion of restrained breeders would shrink, eventually to nothing. A heritable mutation for selfish breeding in the original population would produce exactly the same effect, so even in a completely isolated population, a mutation that eliminated just one individual's restraint on uncontrolled breeding would have the same calamitous result.

Under these circumstances, group selection could only work with the biological equivalent of a police state enforcing a strict ban on immigration, as well as the expulsion or destruction of any individuals caught behaving in a self-interested manner. In the end, even these safeguards would be inadequate to stem the inexorable tide of biological self-interest. Nepotism, the bias of organisms toward their relatives, would ineluctably undermine the group-selected order. Plato, who was far more aware of the significance of kin bias than some contemporary advocates of group selection, thought that unchecked nepotism would corrupt the totalitarian utopia described in his *Republic*. His unworkable and inhumane solution was to stipulate that the state must prevent its guardians from knowing who their relatives are. As we will soon see, even if kin favoritism could be extinguished through some miraculous genetic gimmick, the totalitarian solution would favor the evolution of forms of deception that conceal rampant self-interest behind ostensibly group-directed behavior.[9]

A real solution to the problem of altruism had to wait for an introverted British graduate student named William Hamilton, who accomplished this feat in the 1960s. The idiosyncratic Hamilton was fascinated with the genetics of altruism at a time when nobody else seemed to consider this a worthwhile endeavor, and he often suffered from grave doubts about the value of his work. "At times," he reflected, "I was sure I had seen something that others had not seen. . . . At others I felt equally certain that I must be a crank."[10] Hamilton's paper on "The genetical evolution of social behavior" revolutionized evolutionary biology and became one of the most frequently cited works in the neo-Darwinian canon. It is no exaggeration to say that work on the evolution of social behavior falls into pre-Hamilton and post-Hamilton eras.

Hamilton realized that the reproductive success of organisms is not what drives evolution: it is the reproductive success of their genes. Although these two statements may seem to boil down to the same thing, they are actually distinct. Although your body contains many copies of your genes, there are also copies of your genes lodged outside your body. In particular, your immediate family shares a large number of genes with you. Organisms are genetically linked by what is known as a "coefficient of relatedness," which is probability of a gene being shared with another individual through common descent. Among human beings and many other sexually reproducing creatures, each individual shares, on average, half of its genes with its mother and half with its father; so, each of us has a coefficient of relatedness of .5 with each parent. Using the same reasoning, it is easy to see that we have a coefficient of relatedness of .5 with our siblings, .25 with half-siblings, .25 with each of our grandparents, .125 with first cousins, and so on down the line.

Once we bring the coefficient of relatedness into the picture, it is clear to see that altruistic behavior toward close relatives makes biological sense. Helping family members helps the genes that we share with them: *our* genes. Sometimes, it is more biologically advantageous to concentrate one's efforts on enhancing the fitness of a relative than on trying to produce offspring of one's own. A woman who pours energy into caring for her grandchild thereby enhances the reproductive prospects of 25 percent of her genes. Given that she is likely to be nearing the end of her reproductive life, if not already past it, this may be her best option. A mother performing the same action benefits a full 50 percent of her genes. Hamilton demonstrated that apparently selfless actions might be deeply self-interested on the microscopic, genetic level if the likely fitness benefits exceed their costs. The altruism displayed by ants and dwarf mongooses are examples of what Hamilton called kin altruism; in each case the altruistic individual enhances its own fitness by looking after the interests of its nearest and dearest. If not our brother's keeper, then we are at least the keeper of our brother's genes.[11]

What explains altruism between unrelated individuals? Enter Bob Trivers. Born in 1943 in Washington, D.C., Trivers entered Harvard with the intention of studying mathematics and preparing for a career as a civil rights lawyer. A nervous breakdown in his junior year altered these plans forever. In an effort to help, the noted psychologist Jerome Bruner arranged for Trivers to coauthor a book on animal behavior for fifth graders. This plan was more successful than Bruner could have possibly anticipated: Trivers became a biologist.

Trivers tackled the problem of altruism between nonrelatives in his classic 1971 paper "The evolution of reciprocal altruism." He argued that nonkin altruism would be naturally selected if altruistic actions are reciprocated and that the benefits accrued from being a recipient of an altruistic act exceed the costs of performing such actions for others.[12] A little experiment will show how this model works. Imagine a community located in a locale where there is plentiful quicksand. As the inhabitants go about their daily activities, they are in danger of falling into quicksand pools. Further, imagine that anyone unfortunate enough to step into quicksand has only a 50 percent chance of surviving unless someone rescues them, and that the rescuer has only a 5 percent chance of dying in the process. Finally, imagine that the energy costs of rescuing a comrade are trivial compared with the survival probabilities. If the entire population faces the risk, it follows that natural selection will favor those individuals who are inclined to rescue one another rather than those who callously allow one another to sink. In other words, even though altruistic acts such as saving a nonrelative pose an immediate threat to survival, being heroic racks up big advantages in the long run if one is interacting with other altruists. The 5 percent chance of dying is dwarfed by the benefits flowing from being rescued, should the need arise. It is to everyone's benefit to accept the relatively small risk involved in rescue as an insurance policy against the much greater risk of falling into quicksand oneself. Trivers labeled this "reciprocal altruism."

Reciprocal altruism is a human universal, expressed by the maxim that "one good turn deserves another" and embodied in our systems of social interaction, trade, and economics. Human beings are reciprocal altruists because we evolved in conditions that strongly favored its emergence. Members of our species have a long life span (lots of opportunities to return each other's altruistic favors), a low dispersal rate (regular interaction with the same individuals), small mutually dependent and stable social groups (reliance on one another), and a long period of parental care (protracted contact with close kin).[13]

Biologists have observed many examples of reciprocal altruism in nonhuman species. Gerald Wilkinson's fieldwork in Costa Rica on the food-sharing behavior of vampire bats (contrary to the opinion in Hollywood, vampire bats are native to South and Central America, not Transylvania), which was published shortly after Trivers's theoretical paper, provides a classic illustration. The vampire bat (*Desmodus rotundus*) is a tiny rodent that flies off each night to gorge on the blood of large mammals, usually sleeping cattle and horses. Alighting on its prey, the bat makes a tiny incision and introduces its saliva to the wound. The saliva contains an anticoagulant, which keeps the banquet flowing, and a local anesthetic, to keep the source of dinner asleep and oblivious to the feast occurring at its expense. Vampire bats have such voracious appetites that they can lap up 100 percent of their body weight in blood each night, but they starve to death after only sixty hours without food. However, their grisly nocturnal forays are not always successful, and the bats—particularly juveniles who have not mastered the art of the painless bite—often return to the roost with empty stomachs. Vampire bats live in communities in which individuals strike up friendships with one another, and when a bat begins to become critically depleted, his or her roostmates offer first aid by regurgitating blood for their comrade to devour.[14]

The Arms Race

Although altruism can be highly beneficial to both giver and recipient, it is also risky. Sometimes the other party will refuse to reciprocate, or refuse to reciprocate fully, so altruists need to be able to tell the difference between a fair exchange and a swindle. Those unable to do so would be vulnerable to exploitation and therefore handicapped in the struggle for reproductive success. A successful altruist also has to be able to translate this awareness into action: to refuse to have further dealings with defaulters, or take other punitive measures against them. It is worth noting in passing that this implies that cooperative group life, with its ever-present dangers of undercover exploitation, pushes the mind toward increasingly sophisticated forms of social intelligence. We will explore this much more deeply in the pages to follow.

Human beings have what it takes to be effective reciprocal altruists, but what about other species? To investigate this, Sarah Brosnan and Frans de Waal of the Yerkes National Primate Research Center at Emory University in Atlanta trained pairs of brown capuchin monkeys to exchange tokens for food. Normally, monkeys were happy to exchange a token for some cucumber. However, if one monkey witnessed the other being rewarded with a sweet, juicy grape instead of the more prosaic cucumber chunk, or observed its partner getting a treat without having to surrender her token, the primate rebelled. In about four out of five cases, the deprived monkey either refused to surrender its token, or petulantly hurled either the token or the piece of cucumber out of its cage.[15]

Successful "cheating," as biologists call it, allows the perpetrator to make a profit from what would otherwise be a straight exchange. If you lend me a large sum of money and I never repay you, I gain and you lose. However, after having had this experience, you will be unlikely to trust me again, which means that I may have ruined my chances of cooperating with you to our mutual advantage, much less of exploiting you further. Worse, the offended party can alert other members of the community and give the cheater a bad reputation; so, unless the rewards are very substantial, the overt cheater is likely to lose in the end. The ideal solution to this problem is to use one's victims in such a subtle way that they never realize that they are being used. The more effectively the cheater can fool the "sucker," the more effectively the sucker can be manipulated into furthering the cheater's interests. Even in one-off cases, cheaters must craftily conceal their true intentions in order to get prospective victims to play ball. This is why cheating usually happens behind a smokescreen of deception.

The examples given in chapter 2 suggest that our propensity for cheating and deception is an evolved characteristic, honed by natural selection and rooted, as is all evolution, in the differential replication of successful alleles. Genes build organisms prone to deception in order to further their own reproduction. The mirror orchid does not *intentionally* generate wasp-pornography in order to seduce wasps into fertilizing it; the orchid has evolved to produce flowers that have this effect on its gullible pollinators. In our own species, natural selection has favored the evolution of a knack for inauthenticity; we are able to simulate friendship, guilt, sympathy, and other interpersonal attitudes to self-servingly manipulate the behavior of others.[16] Starting with the commonplace examples of clothing, makeup, and coiffures to disguise our appearance, progressing to the small deceptions of social life known as "tact," and moving on through cosmetic surgery to infidelity, criminal fraud, and political propaganda, it is evident that our social lives are drenched with deceit. We do not need science to tell us that we are natural-born liars, but we need it to help us understand what drives our ubiquitous dishonesty, including our puzzling tendency to lie to *ourselves*.

It was not possible to understand the roots of self-deception before the advent of sociobiology. Freud's theory of the mechanisms of defense, the idea that we exclude information about our own desires from the conscious mind to avoid becoming aware of distressing psychological conflicts, was probably the most influential explanation of self-deception ever proposed.[17] Oversimplified versions of the Freudian account eventually gained such a high degree of popularity among both clinicians and the lay public that it became, in effect, the only game in town. Freud's thesis is difficult to square with an evolutionary biological perspective on the human animal. Darwinian theory states that traits selected to become part of human nature must be fitness enhancing, but how can the tendency to keep oneself unaware of one's own instinctual urges possibly enhance reproductive success? Although ignorance can be bliss, this euphoria comes at an exorbitant price. In the Freudian account, self-deception seems self-defeating and biologically disadvantageous.

What might evolutionary biology have to say about self-deception? Bob Trivers and Richard Alexander simultaneously hit on a potent, biologically realistic alternative to the Freudian line that shows how remaining in the dark about one's true motives can actually be beneficial. To set the stage for it, we must first delve a little more deeply into the biological roots of human mendacity. I pointed out in chapter 2 that effective mind reading is the best defense against manipulation. Putting the matter a little more strongly, the propensity to deceive should select for counter-deceptive mechanisms. The logic is straightforward: in a world of liars, it is advantageous to possess a lie detector. In a treacherous social world—one rife with deceit and double-dealing—an individual who is good at detecting dishonesty will be far less likely to be exploited than less skeptical individuals. He or she will be likely to survive longer and reproduce more successfully than others, and thereby pass the deception-detecting gene on to the next generation. Given enough time, the lie-detection allele will spread through the entire population, resulting in a race of good liars with a refined sensitivity to indicators of dishonesty. Once this happens, the goalposts shift. Those individuals blessed with superior deceptive skill, who use tactics that are so sophisticated and insidious that they fly under the radar of the average lie-detecting mind, will gain the upper hand in the struggle for survival. This hegemony will last only until these ultra-deceptive tactics are confronted by individuals who have evolved even more powerful cognitive equipment in a spiraling evolutionary arms race.

The arms race between deception and detection has huge implications for the evolution of human intelligence. Nick Humphrey, professor of psychology in the Center for Philosophy of the Natural and Social Sciences at the London School of Economics, was perhaps the first investigator to grasp the relationship between social complexity and cognitive evolution. In his pioneering 1976 paper on "The social function of intellect," he pointed out that "Like chess, a social interaction is typically a *trans*action between social partners. . . ."

One animal may, for instance, wish by his own behavior to change the behavior of another; but since the second animal is himself reactive and intelligent the interaction soon becomes a two-way argument where each "player" must be ready to change his tactics—and maybe his goals—as the game proceeds. Thus, over and above the cognitive skills which are required merely to perceive the current state of play (and they may be considerable), the social gamesman like the chess player must be capable of a special sort of social planning. Given that each move in the game may call forth several alternative responses from the other player this forward planning will take the form of a decision tree, having its root in the current situation and growing branches corresponding to the moves considered in looking ahead from there at different possibilities. It asks for a level of intelligence that is, I submit, unparalleled in any other sphere of living. There may be, of course, strong and weak players—yet, as master or novice, we and most other members of complex primate societies have been in this game since we were babies.[18]

Primatologists Richard Byrne and Andrew Whiten expanded Humphrey's crucial insight into what is known as the "Machiavellian intelligence" hypothesis. After Byrne stumbled upon the baboon melodrama described in the opening paragraph of chapter 1, in which the juvenile Paul tricked adult female Mel into relinguishing her hard-earned corm, he began to assemble a portfolio of primate disingenuousness. The steadily accumulating evidence convinced Byrne and Whiten that primate deceit is far more widespread than anybody had suspected. These facts suggested that double-dealing and suspicion might have been the driving forces behind the explosion of brainpower that emerged in monkeys and apes. Our direct ancestors, the early hominids, made their debut just one step further along this same evolutionary trajectory, suggesting that the awesome mental powers of the human animal were also cut from Machiavellian cloth. According to this aptly named Machiavellian intelligence hypothesis, our ancestors evolved their high intelligence in response to the increasingly intricate maneuvers of Paleolithic social life.

Robin Dunbar, professor of evolutionary psychology at the University of Liverpool, came to similar conclusions. He found a direct correlation in social primates between the size of the neocortex (the thinking part of the brain) and the group size typical of their species. Dunbar inferred from this that intellectual power evolved as a function of the demands of social life. The cognitive sophistication of our primate ancestors interacted with expanding group sizes to produce potent selection pressure that accelerated the growth of the human mind. Provided there are sufficient resources to feed everyone, expanding group size brings certain advantages. A larger group provides better protection from predators as well as from other human groups. Jane Goodall's observations of wild chimpanzees in Tanzania show that chimps engage in a primitive form of warfare. Goodall looked on while the members of one troop systematically slaughtered the males of a neighboring clan. The fact that chimpanzees and human beings are the only two species that behave in this manner strongly suggests that our common ancestor also had a penchant for inter-group violence, and that this, in turn, was shared by our Stone Age forebears. Archeological and, more recently, genetic evidence point to widespread prehistoric cannibalism. *Homo homini lupus:* man is a wolf to man.[19]

The steady expansion of the size of prehistoric communities had powerful effects because a linear increase in the size of a group results in an exponential increase in its social complexity. In a group with only five members, each person has ten one-to-one relationships to monitor: their relationship with each of the other four members and the other group members' six relationships with one another. But, Dunbar notes, when the group is expanded to twenty, each member has nineteen relationships to worry about, and the number of one-to-one relationships between third parties balloons to 171.[20] If this sounds daunting, you will not be reassured to know that Dunbar severely underestimates the intricacies involved, because he only takes into account relationships between *individuals*. Our social life is not solely based on one-on-one interactions; it also involves relationships with coalitions of individuals. When we add up all of the possible ways the pie can be cut, we end up with a truly staggering number of combinations. In a group of only twenty individuals, there are 1,048,557 relationship permutations!

It may be that it was primates' need to monitor the mind-boggling complexity of their social relations that provided the impetus for the evolution of basic mathematical skills, and that numeracy began as a way to track clique membership. Empirical research shows that basic mathematical operations are hardwired into the human nervous system, and that even preverbal infants and nonhuman primates are able to perform simple calculations.[21]

The Pinnochio Problem

George Steiner, the distinguished literary critic quoted in chapter 1, noted that "The human capacity to lie . . . stands at the heart of speech."[22] Notwithstanding the Machiavellian finesse displayed by nonhuman primates, and presumably by our hominid ancestors, there can be little doubt the evolution of language vastly extended our deceptive repertoire. Mere words make few demands on the speaker. They require no commitment and only a minimal expenditure of energy. It is all too easy to make promises without the slightest intention of following through on them. Words, therefore, are only credible when backed up by biologically costly actions or other signs of honest engagement.

Given that language lends itself to dishonesty, it makes sense to privilege nonverbal signs over mere words when trying to spot deceit. The anthropologist Gregory Bateson pointed out years ago that inadvertent bodily movements, facial expressions, hesitations, changes in the tempo of speech and movement, tone of voice, and irregularities of expression speak volumes about our relationships with others.[23]

When Shakespeare wrote, "False face must hide what the false heart doth know," he may well have underestimated the difficulties.[24] It is extraordinarily difficult to suppress those nonverbal signs that betray our authentic feelings. For example, most of us sense the difference between a genuine and a phony smile. How do we do it? What is it about that cheesy smile that reeks of insincerity? The French neurologist Guillaume Duchenne discovered the answer to this question over a century ago. In his book on *The Mechanisms of Human Facial Expressions,* Duchenne photographically recorded his use of electrodes to contract the individual muscles of a subject's face (actually, an old man suffering from complete facial anesthesia). He noticed that unlike those false smiles that involve only muscles of the mouth, a smile expressing genuine happiness produces contractions around the eyes. In the false, mouth-only "have a nice day" kind of smile the lips are stretched sideways, with no pronounced upward curling of the lips and no laugh lines around the eyes. The counterfeit smile is not learned: it is part of our inborn deceptive repertoire. Even babies reserve the genuine smile for their mothers, and put on a phoney smile for the benefit of strangers.[25]

How do ordinary people, who are not versed in the anatomy of the face, know that these smiles are false? To quote Shakespeare again, ". . . there's no art. To find the mind's construction in the face."[26] We do not have to work it out consciously: we just respond differently to a false smile than we do to a genuinely warm one, without knowing why. In other words, we distinguish these and other nonverbal signals unconsciously, "with an extreme alertness and, one might almost say, in accordance with an elaborate and *secret code* that is written nowhere, known by none, and understood by all."[27]

The contemporary psychologist and expert on nonverbal communication Paul Ekman uses a detailed coding system to identify a range of distinctive nonverbal expressions of emotion expressed by the human face. People in the grip of feelings of disgust involuntarily wrinkle their noses, and people suppressing anger spontaneously tighten their lips. No matter how hard we try to control them, signs of our real emotions seep through. Attending to nonverbal expressions is clearly a potent weapon in the mind reader's arsenal.[28]

Another problem for the would-be liar is the stress produced by deceit. Effective deception is not always easy, especially when the perpetrator has to face a skeptical audience that is prepared to penalize dishonesty. Deception makes us anxious because lying can be dangerous business. At best, being exposed may mean that others will keep the liar at arm's length. At worst, it may be lethal. Under this kind of pressure, even the most determined con artist is likely to get the jitters. Consequently, human liars tend to follow the example of Pinocchio and rat on themselves by involuntary, nonverbal signs.

When we discover that someone is trying to defraud us, our first response is often a surge of rage. There are a few examples of behavior that look like righteous anger in the literature on nonhuman species. Harris's sparrows (*Zonotrichia querula*) are small, attractive birds that congregate in large flocks during the winter months. Some individuals sport darker head and throat markings than the others: these are known as "dominance badges," veritable black belts of the sparrow world. Any bird with a dominance badge permanently wears a sign that reads, "Don't mess with me." Sievert Rohwer, an ornithologist at the University of Washington, dyed light-colored young birds to make them resemble dominant adults. Sure enough, the newly promoted birds received extremely deferential treatment from the rest of their flockmates. After being cow-towed to a while, the former wimps soon developed an attitude and began actively flaunting it ("clothes maketh the man" and evidently, also the bird). When flockmates realized the ruse, they vigorously punished the weaklings in disguise. When Rohwer bleached dominant birds, they suddenly received little respect from their peers. These individuals reacted to their sudden loss of status by becoming hyperaggressive, as though to punish those sparrows that were breaking the rules of the sparrow caste system.[29]

Monkeys do something similar. The tiny island of Caya Santiago off the southeastern coast of Puerto Rico boasts a colony of about 900 rhesus monkeys (*Macaca mulatta*), descendants of 400 monkeys transplanted from India in 1938 and studied continuously since then. Primatologist Marc Hauser noticed that when rhesus monkeys discover food (either "natural" food or provisions of "monkey chow" thoughtfully supplied by scientists), they first scan their surroundings and often give a "food call" summoning others to the dinner table. However, occasionally a monkey chancing upon a windfall selfishly keeps quiet about the discovery. If the other monkeys catch the delinquent in the act, it is "aggressively attacked and injured" by members of the community that it was trying to defraud.[30] A monkey who eats instead of calling gets more food than one who calls before eating, so it is in the interest of each individual monkey to cheat the community. Similarly, it is in each individual's interest to discourage freeloading. Monkeys on the verge of giving in to their baser instincts show telltale signs of apprehensiveness. Just before scoffing down the food, they survey their surroundings to make sure that nobody is looking.

The mirror orchid does not have worry about getting caught, fielding potentially exposing questions, or hiding its shifty eyes behind dark sunglasses, because its deceptiveness is rooted in morphology and physiology rather than social gamesmanship. It does not toil to conceal its intentions. The nearer we get on the evolutionary tree to *Homo sapiens,* the more intraspecific deception becomes consistently tactical and flexible, and the riskier it becomes.

Our impressive skill at thinking several moves ahead is a mixed blessing, for it makes us painfully conscious of the consequences awaiting those cheaters who trip up. The greater the risk, the more self-conscious we become, and heightened consciousness creates a brand new problem. We become nervous liars, and the more nervous we become, the more likely we are to betray our dishonesty accidentally. In an effort to quell the rising tide of anxiety, liars may automatically raise the pitch of their voices, blush, perspire, scratch their noses, or make small movements with their feet as though barely suppressing an impulse to flee. Alternatively, they may rigidly control their voices, suppress any telltale stray movements, and raise suspicion by their conspicuously wooden demeanors. Either way, our bodies seem to sabotage our minds' best efforts at deceit.[31]

If we could selfishly manipulate others while remaining blissfully innocent of our own true intentions, this would go a long way toward solving the Pinocchio problem. People who do not know that they are lying do not need to worry about being exposed, because they have no inkling that there is anything to be exposed. Trivers proposed that self-deception evolved in just this way: the liars' tendencies to betray themselves inadvertently acted as a selection pressure for the evolution of self-deception. Self-deception did not appear in the mental repertoire of our hominid ancestors to protect them from distress *qua* distress, as champions of the mental health industry assume. Instead, it emerged as a tool for social manipulation. "Biologists," writes Trivers, "propose that the overriding function of self-deception is the more fluid deception of *others. . . .*"

> *That is, hiding aspects of reality from the conscious mind also hides these aspects more deeply from others. An unconscious deceiver is not expected to show signs of the stress associated with consciously trying to perpetrate deception.*[32]

As soon as our forebears acquired the ability to deceive one another self-consciously and to anticipate the possibility of deception by one another, the cognitive ground was prepared for the appearance of self-deception. Self-deception helps us ensnare others more efficiently. It enables us to lie sincerely, to lie without knowing that we are lying. There is no longer any need to put on an act, to *pretend* that we are telling the truth. A self-deceived person is *actually* telling the truth to the best of his or her knowledge. Believing one's own story makes it all the more persuasive.

Although there have probably always been unconscious mental states, the capacity for self-deceptive thoughts led our ancestor to actively and intentionally, albeit unconsciously, keep certain thoughts out of the spotlight of conscious awareness.[33] A portion of the mind became specialized for generating and nurturing falsehoods maintained side-by-side with an unconscious grasp of reality. The capacity for language must have vastly enhanced the capacity for self-deception. Language allowed us to lie (in the narrow sense of the word) to others and to whisper self-serving falsehoods to ourselves. A portion of the brain developed special expertise in dishonesty, cleverly weaving useful illusions out of biased perceptions, tendentious memories, and fallacious logic.[34] I will have more to say about the crucial impact of language acquisition in chapter 5.

Consequently, modern human beings are naïve realists who take for granted the accuracy of their misrepresentations of the social world, but actually systematically misconstrue both themselves and other people.[35] Social psychologists have long known that a good deal of what we consider "objective reality" is actually the product of our "unnoticed interpretative manipulations."[36] Before evolutionary psychology, it was not possible to understand exactly what drives people to distort their perceptions, memories, and logic in this way. A biological perspective helps us understand why the mental gears of self-deception engage so smoothly and silently, imperceptibly embroiling us in performances that are so skillfully crafted that they give every indication of complete sincerity, even to the performers themselves.

If self-deception is so useful, why is it that we do not deceive ourselves all the time? First, deception—and therefore self-deception—is not always advantageous. The human community contains both conflicts and confluences of interest. Second, we have to be somewhat more economical with lies than we are with the truth: lying all the time would be self-defeating. Aesop's fable about the boy who cried wolf is an excellent lesson about what biologists call the "frequency-dependent" nature of deception. Someone who tried to deceive others all the time would not be a credible liar. Given that the biological function of self-deception is to deceive others, and that it is not advantageous to try to deceive others all the time, there is no point engaging in self-deception around the clock. Third, even the judicious use of self-deception has a price tag attached: it helps us help ourselves to ill-gotten social gains, but it also deprives the conscious mind of potentially useful information. Unless self-deception is limited to the right dosage, the disadvantages of information deprivation would outweigh the benefits of social manipulation and nature would select it out of existence.

The propensity to lie to oneself is frequency dependent, relatively fluid, and exquisitely sensitive to the nuances of the social situations in which we find ourselves. It catches us off-guard, swinging into action when we least expect it. Is the idea of a mind that is not aware of its own activities credible; and, if so, what are the mechanics of these peculiar processes? To answer these questions we will temporarily have to leave evolutionary biology behind and turn to the field of cognitive science in search of answers.

4

The Architecture of the Machiavellian Mind

Our mind is so fortunately equipped that it brings up the most important bases for our thoughts without our having the least knowledge of this work of elaboration. Only the results become conscious. This unconscious mind is for us like an unknown being who creates and produces for us and finally throws the ripe fruits into our laps.

—Wilhelm Wundt

If self-deception took root in the human mind because it enabled us to be better deceivers by protecting us from clever mind readers, this must have significant impact on how human minds work. Trivers's concept of the adaptive role of self-deception is that there is a deep fissure running through our inner landscape, dividing it into conscious and unconscious regions. In this chapter we will turn to cognitive science to assess this picture and then move on to consider some of its further ramifications.

Science and the Unconscious Mind

The Machiavellian mind is a partitioned mind with one sector open to the public and the other kept deeply private, so private, in fact, that even its owner does not suspect its existence.[1] In short, there must exist an unconscious mind lying beyond the reach of introspection. Surprising as this may sound, scientists have been accumulating evidence for decades that demonstrates that introspection gives us only a partial and misleading picture of our inner lives. The scientific investigation of unconscious mental processes presents us with a map of the inner landscape far removed from reassuring commonsense views.

Although we cannot legitimately draw scientific conclusions from subjective experiences, personal experience provides a point of contact with the strange domain that we are now setting out to explore. I will therefore begin with some phenomena that we have all experienced. After we get our bearings, we can grapple with the empirical work of the psychologists.

All of us have had ideas "pop" into our heads, unexpectedly and seemingly at random. Thoughts sometimes seem to *happen* to us, tumbling into our consciousness like uninvited, although not necessarily unwelcome, guests. While doing nothing in particular, you find yourself mentally singing a song, thinking of an old friend, or savoring the taste of something you enjoy eating, say a bagel with lox and cream cheese. A great deal of our normal stream of consciousness has this peculiarly passive, apparently automatic, quality. Pause for a few minutes to observe faithfully the path taken by your thoughts from one moment to the next. You will notice that they have the weird discontinuity of a dream.

Real dreaming provides a particularly forceful example of intrusive thought. We are not the conscious authors of our dreams: they come to us unbidden, like nocturnal apparitions. The German language has a word for unwilled thoughts: *Einfalle*. When Freud instructed his patients to recline on the psychoanalytic couch and "free associate," he actually told them to engage in *freier Einfall*, that is, to observe and report the thoughts freely intruding into their consciousness. Freud thought that *freier Einfall* provided a method for accessing the unconscious mind. As we will see, he was not all that far from the truth.

Many creative people in the sciences and the humanities have described drawing on involuntary thought in their work (for examples, see Appendix I). The celebrated French mathematician/physicist Henri Poincaré chronicled a particularly impressive example of unconscious scientific creativity in a famous paper on "Mathematical creation."[2] Poincaré worked hard for more than two weeks to prove that a group of mathematical functions called "Fuchsian functions" could not exist. One evening, after a long day wrestling with the problem, he drank some black coffee before going to bed and had difficulty falling asleep. Tossing and turning, he had visual images of ideas emerging, interlocking, and forming pairs. When he woke up the next morning, he *knew* that Fuchsian functions existed.

Next, Poincaré left his home in Caen to go on a geological expedition. Just as he stepped into a bus in the town of Coustances, the idea suddenly struck him that the transformations he had used to define Fuchsian functions were identical to those used in non-Euclidean geometry. Upon returning to Caen, he took out pencil and paper and proved it. Poincaré then turned his attention to other problems in arithmetic, problems that, as far as he was aware, had no connection at all to Fuchsian functions. Reaching an impasse, he decided to go on a short vacation. While ambling along some cliffs by the sea, a new revelation came to him like the proverbial bolt from the blue "with . . . brevity, suddenness and immediate certainty."[3] He returned to Caen and found that his brainwave had revealed an entirely new class of Fuchsian functions.

Working systematically through the problems exposed by the new discovery, one conundrum remained that Poincaré could not manage to solve. Although his usual strategy in this kind of situation was to temporarily abandon conscious work on the problem and let his unconscious mind take charge,[4] on this occasion life intervened: he was drafted into the army. It was during his stint in the military that the solution "suddenly appeared" to him one day as he was walking along the street.

Considering Poincaré's weird experiences, it is easy to sympathize with the ancients who believed that sudden insights were actually messages from supernatural beings. Even Socrates believed in his *daimonon,* a sort of guardian angel who mentally conversed with him. In our own era, the Indian mathematical prodigy Srinivasa Ramanujan made mathematical discoveries in dreams that he believed to be messages from the Hindu goddess Namagiri.[5] Today, in the Western world at least, we are less inclined to explain events such as these as supernatural visitations than we are to attribute them to "intuition." However, explaining Poincaré's discoveries as intuition tells us nothing. Indeed, the notion of intuition, taken on its own, seems only marginally less mysterious than the Socratic guardian angel theory. If the concept of intuition is going to do any explanatory work, we must harness it to a scientific account of how the mind operates.[6] Poincaré provides a clue about how to proceed.

> *Often when one works at a hard question, nothing good is accomplished at the first attack. Then one takes a rest, longer or shorter, and sits down anew to the work. During the first half-hour, as before, nothing is found, and then all of a sudden the decisive idea presents itself to the mind.*[7]

This is a common experience. Working on a problem and getting nowhere fast, you decide to take a break. After your breather, you return to the table and quickly arrive at a solution. What is going on here? Perhaps your conscious mind, refreshed by the break, is able to work more efficiently. This is possible, but Poincaré suggested a very different explanation. He thought that when we take a break, when we stop consciously focusing on a problem, the unconscious mind has an opportunity to work on the problem without the interference caused by conscious effort.[8] According to Poincaré, in these scenarios it is not the conscious mind that solves the problem, although it likes to take the credit for doing so.

Poincaré was not alone in turning to the unconscious solutions to scientific problems. Thomas Edison is reputed to have worked this general concept into an ingenious technique, a creativity-pump. When he reached an impasse in his work, Edison took a catnap . . . with a difference! He seated himself in a comfortable chair and nodded off while holding steel balls in his hands. As soon as he fell asleep, his hands relaxed and the balls dropped into pans he had placed on the floor for this purpose. The din created by the balls hitting the pans woke him up, often with a new idea about how to carry his project forward. (Edison was extremely hard of hearing, so this exercise was less harsh than one might imagine.)[9] Albert Einstein was also devoted to creative napping. He described the same phenomenon as follows:

> One phenomenon is certain and I can vouch for its absolute certainty: the sudden and immediate appearance of a solution at the very moment of sudden awakening. On being very abruptly awakened by an external noise, a solution long searched for appeared to me at once without the slightest instance of reflection on my part—the fact was remarkable enough to have struck me unforgettably—and in quite a different direction from any of those which I had previously tried to follow.[10]

It is easy to miss the significance of Poincaré's careful wording: the answer *presents itself* to awareness. Similarly, Einstein describes a solution *appearing* to him. On these occasions, the conscious mind is a relatively passive recipient of insight. It is common for scientific discoverers to describe their moments of epiphany in such terms. "I can remember the very spot in the road," wrote Charles Darwin in his autobiography, "whilst in my carriage, when to my joy the solution occurred to me."[11] Darwin had struggled to find an answer to the question of how evolution works, but in the end, the answer seemed to find *him* (as we say, the answer had previously *eluded* him). Andrew Wiles's account of solving Fermat's last theorem has the same sense of cognitive intrusion: "suddenly, totally unexpectedly, I had this incredible revelation."[12] In ordinary conversation we are liable to say that a realization "hit us," sometimes with "blinding force." We often feel that insight penetrates our consciousness from the outside.

Think about the times that you have searched for and eventually found a misplaced item (say, a set of car keys). You know that you put it somewhere, but you cannot remember where. After an hour or two of increasingly desperate searching for the keys, punctuated by frustrating attempts to recall when and where you last laid eyes on them, you sit down forlornly, almost reconciled to the prospect of failure, when all of a sudden you just *know* that the keys are in the pocket of the jacket that you wore yesterday afternoon. For an even more mundane example, perform the following mini-experiment. Think of your date of birth. The correct date and year immediately occurred to you. Now, how did you retrieve this information? Can you spell out the procedure systematically, in the same way that you can describe the sequence of actions required to make French toast? You are unable to do this because you have no inkling of the mental processes involved. The whole operation was rather like using a search engine to find information on the Web. You typed in the key words "my birthday," clicked on "search," and waited for the results to appear on the screen of consciousness.

Many unconscious processes structure and inform our daily lives. They are so ordinary that we take them for granted. How do we manage to express our thoughts effortlessly in grammatically correct sentences, delicately coordinating mouth, tongue, and larynx to pronounce the words? A mere minute of speech involves ten to fifteen *thousand* neuromuscular events.[13] We are not conscious of the symphony of precise manipulations required of our vocal apparatus for even the simplest verbal interchange. The words just seem to spill out of us, accompanied by only a vague sense of the grosser movements of mouth and tongue. How do we learn our mother tongue in the first place? What mental procedure makes it possible for you to read these sentences, instantly translating complex sequences of black marks on paper into meaningful propositions? How does anyone manage to quickly perform the overwhelmingly complicated computations required to catch a baseball or to drive the car to the corner store? In fact, how does anyone learn anything at all? Introspection provides us with no information about how we perform basic cognitive tasks. They lie in that hidden region of the mind that the famous developmental psychologist Jean Piaget dubbed the cognitive unconscious.[14]

During the late nineteenth and early twentieth centuries, psychologists, neuroscientists, and philosophers were beginning to take the notion of unconscious cognition seriously, but its scientific legitimacy took a nosedive when the psychological movement known as behaviorism became dominant. Until the behaviorists came along, the focus of psychology was to investigate mental phenomena. The behaviorists redefined psychology as the science of the prediction and control of behavior, and considered "mentalistic" notions such as the unconscious mind to be insipid and unworthy of scientific attention. After four decades, the emerging discipline of cognitive science broke the behaviorist stranglehold on the science of the mind. The effect of the "cognitive revolution," as it is often called, was like dismantling an intellectual Berlin Wall. Scientists gradually came to realize that if they were to make sense of the human mind, they would have to give the unconscious a seat at the table. As research progressed, they became accustomed to the idea that the driving forces of the mind lie deep within its bowels, beneath the superficial and misleading level of conscious awareness. Psychologists were at first rather uneasy about using the word "unconscious." They recoiled from the dense tangle of "unscientific" Freudian connotations it had accumulated, and grasped at euphemisms such as "implicit," "tacit," "incidental," "without awareness," "nonconscious," and "automatic" to avoid accusations of keeping bad company. Arthur Reber, a cognitive scientist at Brooklyn College well known for his work on unconscious cognition, candidly recalls:

> *Back in the 1960s when we began working on the general problem of implicit learning, I felt a distinct reluctance to use the term* unconscious *to characterize the phenomena appearing with regularity in the laboratory despite the obvious fact that the processes we were examining were just that. There was (and to some extent, still is) considerable semantic spread from the psychoanalytic community's use of the term; for a young researcher in the as yet unacknowledged cognitive revolution, any suggestion of conceptual familiarity here was seen as seriously compromising to one's stature.*[15]

By 1987, the intellectual climate had changed so dramatically that the psychologist John Kihlstrom could unblushingly write in a famous article entitled "The cognitive unconscious" that even higher mental processes are performed without the benefit of conscious awareness. The unconscious had finally come out of the closet.

Unconscious Computations

The invention of the digital computer in the 1940s and the emergence of artificial intelligence shortly afterwards nurtured the sparks of the cognitive revolution. The developing notion that the brain is a flesh-and-blood computer, a neural information-processing device, turned out to be an immensely powerful metaphor for how the mind works.

The computational theory of the mind was not devoid of controversy. Conflict soon developed between two broad views of the relationship between minds, brains, and computers. Supporters of Strong Artificial Intelligence claim that the chunk of meat between your ears literally *is* a computer, and that computers are able to think. It doesn't matter that brains are warm, squishy, and slimy whereas artificial computers have the cold, hard lines of metal, plastic, and silicon: what counts is what brains and computers *do*, rather than the materials from which they are built. On this functionalist view, the densely interconnected neurons that transmit electrical signals through the brain are equivalent to the logic gates that regulate the flow of electrical pulses through the processing unit of a PC.

The proponents of Weak Artificial Intelligence reject this thesis. They argue that computers only simulate thinking. While the details of this long-standing debate need not concern us here, what is important for the purposes of this discussion is that both versions of AI made it easier for scientists to entertain the idea that mental processes do not need to be conscious. If computers think, or simulate thinking, then surely they do this unconsciously. Not even the most ardent advocates of Strong AI believe that machines currently exist that have experiences, subjective inner lives, and so on, although some advocates of Strong AI are theoretically committed to the possibility that this lies in the future.

To get the flavor of the way that computational theorists pictured unconscious information processing, consider how cognitive scientists have described the infrastructure of the act of vision. From an ordinary commonsense standpoint, vision seems straightforward: we just look at something and we see it, right? Wrong. When it comes to vision, there's an awful lot more going on than meets the eye. The idea that vision depends on unconscious cognition predates contemporary cognitive science. The brilliant German physicist Herman von Helmholz wrote in his 1860 *Treatise on Physiological Optics:*

> *The psychic activities that lead us to infer that there in front of us at a certain place there is a certain object of a certain character, are generally not conscious activities, but unconscious ones. In their result they are the equivalent to* conclusion, *to the extent that the observed action on our senses enables us to form an idea as to the possible cause of this action. . . . But what seems to differentiate them from a conclusion, in the ordinary sense of that word, is that a conclusion is an act of conscious thought. . . . Still it may be permissible to speak of the psychic acts of ordinary perception as* unconscious conclusions. . . .[16]

Helmholz was not in a position to fill in the details. Nearly a century later, British psychologist David Marr provided a detailed computational account. According to Marr, after light enters the eyes, which project an inverted image on the 126 million photosensitive cells of each retinal screen, the retinal cells churn out information about the intensity of light detected at each of these points. Next, the nervous system uses this information to detect sudden jumps in light intensity, interpreting these as edges of and boundaries between objects. This operation is performed by specialized neural modules called "zero-crossing detectors," which also filter out extraneous information or "noise" by calculating the average values for specific regions of the visual field. Once the brain establishes the steepness of these gradients, other processes kick in to distinguish between "bars," "edges," and "blobs," encoding their position, orientation, length, contrast, and width. It is only at this rather advanced stage of visual processing, which occurs independently for each eye, that information detected by the two eyes is brought together and compared. The brain uses several "rules" to map the two slightly different images onto one another to generate a fused representation that eventually becomes a three-dimensional image. The entire process, which is greatly simplified in this summary, takes place in less than a second and is well beyond the reach of introspection. All that we are ever conscious of is the result of this complex sequence of mental operations: the three-dimensional visual image.

Fictive Consciousness

Meanwhile, the University of California cognitive neuroscientist Benjamin Libet was unearthing some unsettling facts about consciousness and free will.[17] Libet was interested in determining just how the brain goes about making decisions to act. He asked his subjects to press a button whenever they wished and to note the time displayed on a rapidly moving clock the very moment that they felt an impulse to press the button.

Libet attached electrodes to monitor activity in his subjects' motor cortex, the part of the brain that controls movement, so that he could measure the electrical tension that mounts as their brains prepared to initiate the action. Common sense would tell us that his subjects' readiness potentials would begin to build as soon as, and no sooner than, they consciously decided to push the button. We are inclined to think this because we assume that it is the conscious decision to push the button that initiates the act of pushing the button. Libet found that this is not what happens. Our brains begin to prepare for action over a third of a second *before* we consciously decide to act. Although we go through life believing that decisions to act begin in consciousness, the experimental evidence suggests that this is a universally shared delusion.[18]

Libet also studied the relationship between sensory stimulation and consciousness by monitoring changes in the electrical activity of the brain caused by incoming information called "evoked potentials."[19] Normally, we feel sensations when our sense organs activate a region of the brain called the sensory cortex. (Although sensations *seem* to occur in various parts of one's body, they are actually "in" the brain.) When an experimenter directly stimulates the sensory cortex with an electrical pulse, its owner experiences a sensation. For example, when the area of the sensory cortex corresponding to the right hand is stimulated, the person will feel a sensation in their right hand, just as if someone had touched that hand. Libet noticed that although even very brief electrical stimulation of the brain results in evoked potentials, no conscious experience results unless the stimulation continues for approximately half a second. In short, we are unconsciously sensitive to sensory inputs that are so ephemeral that they bypass consciousness altogether.

Stranger still, Libet found that there is a 500 millisecond, or one half second, delay between pricking the skin with a pin and consciously feeling the jab. We are normally unconscious of the gap between a stimulus and the sensation it produces, believing that we feel the pinprick at the moment that it occurs. Libet investigated this phenomenon by stimulating the sensory cortex so that his subjects would experience a tingling sensation in their right hand while simultaneously stimulating manually the skin of their left hand. He then asked them which sensation they felt first: the tingling of their right hand, or the sensation in their left. Common sense tells us the sensation in the right hand, the one produced by direct stimulation of the brain, should be the first across the finish line. After all, the stimulation of the left hand has to travel all the way up the sensory nerves before reaching the brain, whereas the direct stimulation of the sensory cortex has a huge head start. By this point, you have probably begun to realize that common sense is a rather poor guide to understanding the mind, and will not be surprised to hear that the commonsense assumption did not prove true. Libet found that even when the brain was stimulated as much as two fifths of a second *before* the skin of the left hand, the subjects reported feeling the stimulation of their left hand first.

This is the neurophysiological equivalent of someone entering a revolving door behind you but coming out ahead of you. Did the nerve impulses from the hand take a fast lane to consciousness, somehow overtaking the impulses coming from the direct stimulation of the brain? Did they manage to transcend space and time? Not at all. It takes about one fifth of a second for the stimulation of the skin to produce evoked potentials in the brain, and an additional one half second for the evoked potential to produce a conscious sensation. This means that we actually experience sensations almost a full second after their stimulus occurs. Why is it, then, that when you touch your left hand with your right (go ahead, try it) you seem to feel the touch at the very moment that one hand meets the other? Why isn't daily life like a badly dubbed foreign movie, with our experiences permanently out of synch with the action? Why does our subjective life appear seamless? Libet proposed that the brain unconsciously papers over the discontinuities, referring sensations backwards in time by seven tenths of a second in much the same way that one attempts to conceal a late payment by predating the check. This ingenious trick makes it *seem* to us that we respond to stimuli instantaneously.

At the most fundamental level, these experiments demonstrate that conscious experience operates with a systematically distorted picture of reality. The brain, it seems, fills in the gaps and smoothes the jagged edges of perception to create fictional continuities. Conscious experience is more dreamlike, more a construction, fiction, or fabrication than most of us would care to believe.

Still not convinced? Ponder the enchantingly named phenomenon of "confabulation across saccades." As we look at our surroundings, our visual experience seems coherent and continuous. In fact, our eyes never keep still for very long, even when we think we are staring steadily at a fixed point in space. Our eyes remain fixed on one spot for no more than a quarter of a second before darting off in tiny, jerky movements called saccades. Although we are effectively blind during saccades, we do not have choppy visual *experiences* because our brains silently, unobtrusively, and unconsciously iron things out for us, editing the visual information to create the illusion of a smoothly unfolding visual panorama. In one of several marvelous experiments demonstrating saccade-blindness, the experimenter asks his or her subject to gaze at a scene projected on the screen of a computer monitor. Of course, this is no ordinary computer. The experimenter has rigged it to change some aspect of the picture during the viewer's saccades. Because the subject is blind during saccades, he or she is completely oblivious to these sometimes-dramatic alterations, even if told beforehand that the picture will change at some point during the experiment.[20]

These and other experiments show quite clearly that the brain has no difficulty scrambling messages before they reach the conscious mind. The manipulation of information before it gets to consciousness makes self-deception possible. There is nothing inherently incredible about the idea that our brains tendentiously manipulate information about our social interactions, just as they manipulate sensory and motor information. The interval between unconscious awareness and conscious perception on one hand, and unconscious volition and the conscious "decision" to act on the other, allows plenty of time for the Machiavellian unconscious to tamper with the information before it reaches consciousness. If we are to have any hope of truly understanding our minds and ourselves, we must find some way to penetrate this illusion, not by introspection, but by scientific investigation.

Consciousness and the Modular Mind

Imagine standing at a cocktail party. The room buzzes with dozens of conversations, but in spite of the din, you are able to hear what the person talking to you is saying. As long as your attention remains glued to your conversational partner, all of the other conversations fade into a background murmur. However, as soon as boredom sets in, or when you become curious about what someone else within earshot is saying, a mere shift of attention enables you to "tune in" to another conversation. The cost of tuning in to another conversation is that you have to tune out of your present one.[21] In short, the cocktail party effect (as psychologists call it) suggests that we are able to follow only one conversation at a time, and never two or more concurrently.

During the 1950s, psychologists hit upon an ingenious way to study this effect. They placed a set of stereophonic headphones on a person's head, played a different message into each ear, and found that only one of the messages, the "attended" message, was consciously understood. The experimental subjects were not totally oblivious to the unattended message.

They were vaguely aware of how it sounded, but could not pick up its meaning (it was rather like listening to a message in a language that one does not understand). Investigators concluded that our brains process the semantic properties of language, the meaning of words and sentences, separately from its acoustic properties. One part of the brain picks up the sound of the speaker's voice, while a different part determines the meaning of the words. The British psychologist Donald Broadbent thought that these experiments showed that information flows into the mind through several parallel channels and that the "meaning" channel has a rather limited capacity. The data seemed to demonstrate that meanings must walk into consciousness single file, that only one message at a time can squeeze through the narrow door that leads semantic information to consciousness. Later research provided a different picture. Although it is true that the meaning of unattended messages never reaches consciousness, a variety of ingenious experiments demonstrated that these messages are nevertheless unconsciously understood, at least somewhat.[22] For example, experimenters presented subjects with an ambiguous sentence in the attended channel, such as "They threw stones at the bank yesterday." This sentence is ambiguous, because the word "bank" might refer either to a financial institution or the bank of a river (and the stone throwers, either anti-capitalist demonstrators or kids testing their marksmanship). At the same time that this sentence enters one ear, a disambiguating word (in this example either "money" or "river") is fed into the other, unattended channel.[23] The investigator then asked them to choose between two rival interpretations of the sentence: either "They threw stones at the side of the river" or "They threw stones at the Savings and Loan Association." Contrary to Broadbent's hypothesis, subjects tended to select the interpretation that was consistent with the disambiguating word, showing that unattended messages are nonetheless unconsciously understood.[24]

These experiments suggest that the conscious can carry only a very limited informational load. Picture the mind as a funnel, with consciousness as the narrow end. No matter how much information pours into the wide end of the funnel, it can only pass through the spout at a slender trickle. The trickle of conscious thought is extraordinarily sluggish. Unconscious acts like choosing words to slot into a sentence, returning a tennis serve, scratching an itch, or slamming on the brakes of your car to avoid an accident are so fast that they are often finished by the time we become conscious of them. An accomplished typist churning out sentences at the rate of 120 words a minute can type an entire word in the 500 milliseconds that Libet has shown it takes for us to consciously respond to a sensory stimulus. As her agile fingers fly over the keyboard, the typist has actually finished a word in less time than it takes her to be consciously aware of the word she is typing.[25] The same principle holds true of our emotional reactions. We often respond emotionally to a situation before we become conscious of our emotional reaction.[26] Anger, for example, comes in a *flash*. Grief or terror *overtakes* us. The fact that our emotions often slip out before we have a chance to realize what is going on is part of what makes them difficult to conceal from others.

Given that a whole range of mental processes can occur without the involvement of consciousness, the obvious question is, "What do we need consciousness for?" There have been numerous suggestions made. At one time or another, psychologists have asserted that consciousness is an essential component of learning novel or complex stimuli and skills, choosing between competing inputs, reflective responding memory, and planning (this list is not exhaustive). However, not one of these proposals is fully consistent with the scientific evidence. The disturbing, and at first glance starkly counterintuitive, answer to the question of what we need consciousness for may well be "very little." To the dismay of many traditionalists, science is gently escorting consciousness to the periphery of the mind.[27]

Freud was probably the first person to provide a detailed psychological model that displaces consciousness from the center of the mind. During the late nineteenth century, when Freud was pursuing a career in neuroscience, it was widely believed that all genuinely mental states are conscious. Freud was aware of the growing body of evidence from neurology, psychiatry, and psychology that contradicted this view, and he grew tired of the intellectual contortions needed to squeeze his clinical observations into the Cartesian straightjacket. So, in the spring of 1895, he set out to rethink his conception of mind and, in particular, his notion of consciousness, setting out his thoughts in a manuscript posthumously published as the *Project for a Scientific Psychology*. Freud's *Project* brilliantly sketched out a neuropsychological theory of mind based on what we now call "connectionist" or "neural network" principles (in fact, the concept of backpropagation, which plays a central role in neural net theory, was derived from Freud's *Project*).[28]

Freud's new vision was radical. He jettisoned the fashionable but ultimately unworkable orthodoxy that mind and body are distinct entities, replacing it with the view that mental processes are the activities of a physical organ, the brain. Freud thought that the brain must be composed of at least three major functional units for it to serve as the vehicle of our mental life. First, information needs to get into it from the outside. The brain must therefore have an input system capable of absorbing incoming information and then quickly bouncing back to its original state to be ready for any new impressions that the world throws at it. Second, the brain must be able to store and analyze the information that it has received. Freud thought that there is a close connection between analysis and storage, because analyzing information involves comparing it to information previously stored in memory. Unlike the input system, this processing/memory system must preserve traces of the modifications that it undergoes, in order to lay down the physiological basis for memory.

So far, the model seems sensible, but hardly revolutionary. Freud's most radical move lay in the way that he conceptualized the output system. We usually think of behavior as output from the mind. So, for example, one might see an object with a certain shape, color, and texture (input), recognize it as a peach (processing), and then reach out to grasp the peach and eat it (output). Freud regarded conscious thought, as well as behavior, as the system's output. This was quite a departure from what was at the time an almost universally held assumption that consciousness is an intrinsic feature of thinking. By detaching thought from consciousness, Freud in effect proposed that *consciousness has no role in mental processing.*

If Freud were alive today, he would doubtless illustrate this conception of the mind with that tried-and-true metaphor, the computer. Here is how the analogy would work: As I type these words, I am inputting information into the personal computer on my desk. Each keystroke is represented in the central processor by a series of electrical pulses. As I gaze at the monitor, I can see the results of my work displayed before my eyes. Freud's input system is analogous to the keyboard attached to the computer, which enables me to create these words and sentences. The processing unit buried deep in the guts of the machine where the real work gets done plays the role of the cognitive/memory system, where information is analyzed and stored. The monitor, which merely displays the results, is the counterpart of consciousness. Like a computer monitor, consciousness neither does the cognitive work nor informs us about what is going on in the depths of the brain (the monitor shows only the results of the CPU's efforts, not how they were obtained). Consciousness provides "neither complete nor trustworthy knowledge" of mental activity.[29]

In their haste to distance themselves from the pseudoscientific excesses of psychoanalysis, many contemporary cognitive scientists like to emphasize the dramatic differences between their conception of the unconscious and Freud's. The Freudian unconscious, they say, is repressed and irrational and has little scientific evidence in its favor. In contrast to this, the cognitive unconscious is not excluded from consciousness: it is unconscious in consequence of the structure of the mind. It is implicated in a range of adaptive behaviors, is backed up by solid empirical research, and is certainly nothing like the wild beast that was supposedly Freud's exclusive concern. This simple dichotomy betrays a gross misunderstanding of Freudian theory. In fact, Freud considered the "repressed" unconscious as a special case of a more general cognitive unconscious. Although generous in identifying his shortcomings, the new explorers of the mind have generally failed to give the pioneer his due.[30]

Although Freud was not—and is not—alone in claiming that *all* thinking is unconscious, this idea remains controversial.[31] Surely, there must be definitive evidence proving that at least sometimes we think consciously! Well, not really. Max Velmans, a cognitive scientist at the University of London, argues that voluminous experimental literature indicates that there is no credible scientific evidence that consciousness plays any role at all in cognitive processing. Consciousness may *depict* a mental process, or display the results of mental processing, but it does not do any of the cognitive work. The belief that we think consciously confuses the display with the real scene of action, like a small child who believes that there are little people inside the television set.[32]

A Community of Demons

Many cognitive scientists understand the mind as composed of functional units called "modules." To get a handle on this idea, imagine that you have moved into a new home that is in need of extensive renovations: the roof needs repairing, there is old plumbing to be replaced, the electrical circuits are shot, and so on. Looking for help, you can either hire a single Jack-of-all-trades to do all of the jobs, or employ a specialist for each task. Whereas some psychologists consider human intelligence to be a cognitive Jack-of-all-trades—a single, general-purpose function (so-called general intelligence)—others, including most evolutionary psychologists, conceive of it as a team of specialists. According to the modularity thesis, our skulls are host to a number of specialist minds, each of which tackles a different kind of adaptive task. These sub-minds are what is meant by the term "cognitive modules."[33] Although some deny the existence of general intelligence that operates across domains, this extreme position is not mandatory. Most evolutionary psychologists accept the existence of general intelligence, but believe that it coexists with the more specialized modular units.[34]

On purely pragmatic grounds, it is better to have a team of experts than a Jack-of-all-trades. Making an analogy with a toolbox, if you often need to tighten leaky pipes, it is a good idea to own a pipe wrench. By the same token, if you (actually, your ancestors) repeatedly encounter certain problems in living that require particular cognitive and behavioral responses, then it is useful to own a mental "tool" that will deal with such situations automatically and efficiently. This is the way that nature designed our bodies: we have special organs for special functions, not just one general-purpose organ. Why should mental adaptations be any different?[35]

Human mating behavior appears to be the function of a specialized cognitive module. Despite undeniable cultural variations, human beings the world over appear to respond in a remarkably uniform manner to members of the opposite sex. One feature to which both men and women automatically respond is the degree of symmetry in a prospective partner's face. Both prefer symmetrical to asymmetrical features, faces in which the left side is almost a perfect mirror image of the right. As it happens, facial symmetry correlates with fertility in women and sperm quality in men. The in-built nature of this preference is driven home by the fact that even small infants prefer looking at symmetrical faces. Now, none of us consciously calculates amount of symmetry in the faces of prospective lovers; an unconscious cognitive module dedicated to mate-selection does the work automatically. The conscious mind simply responds, finding some faces attractive and others unattractive without knowing precisely why. Another, less well-known example is what psychologist Paul Rosin calls the "contagion system." Rosin's work shows that our attitudes toward filth and contamination are not simply the upshot of social learning; they have species-wide features that bear all the earmarks of a specialized, hardwired module. From Toledo to Timbuktu, human beings use special inferential rules for making judgments about risks of contamination, which are distinct from the way they reason about other domains. For instance, we do not consider the degree of risk to be proportional to the degree of contact with a source of contagion. There is no dose-response curve: even a slight contact with, say, a diseased individual is felt to transmit the whole of the risk of contagion.[36] There is no consensus about just how many modules the human mind contains.

Modularity theory, applied to self-deception, suggests that our mental life is carried along by specialized teams of "demons" (to use Oliver Selfridge's colorful term[37]), that may or may not display their information in consciousness. We need to discard the conventional idea of the conscious mind as a powerful executive who is in charge and passes orders down the hierarchy of processing systems to his low-level flunkies. We are better served by a vision of the mind closer to that expounded by Tufts University philosopher Daniel C. Dennett, who argues that our minds harbor a seething crowd of competing versions of reality, all jogging for control. "Mental contents become conscious," he writes, ". . . by winning the competitions against other mental contents for domination in the control of behavior."[38]

Imagine the mind as an immense room brimming with intense activity. There are many tables in the room, around which sit teams of specialists, who analyze streams of data pouring into their computers from the outside world. Some of these groups communicate and cooperate with one another by exchanging information, while others are not even on speaking terms. There is a narrow slot in the wall at the far end of the room, and the only way to publish a report is to pass it through the slot. Every day the experts produce voluminous documents, but the slot is so narrow that they can only pass it through one page at a time. Consequently, only a small fraction of their work is ever made public. In this model, the teams of experts represent the modules and submodules, the community of "demons," that populate the human mind/brain. The information received by their computers is the data that our brains receive from our sense organs. Finally, the unprepossessing mail slot corresponds to consciousness: the narrow end of an informational funnel (later, we will discover that a team of poets helps the scientists prepare their reports for publication). The huge disparity between the massively parallel activity of the modules and the extremely limited representational capacity of consciousness creates an information bottleneck through which thoughts can pass only in single file. This gives us the illusion of consciously thinking one thought after another (so-called serial processing). To shift the metaphor yet again, the unconscious cognitive activity is like a symphony, or a cacophony, whereas consciousness whistles a tune.

If consciousness has any real role in mental activity (and as we have seen, that is a big "if"), there would be no point in overloading it with information that it does not need. If the mind can perform a task unconsciously, it makes sense to leave consciousness free for other tasks. But this is not all there is to self-deception which, as we have seen, demands that certain information about our own motives and intentions be *kept out* of awareness. According to the evolutionary perspective outlined earlier, this knowledge embargo is fitness enhancing, improving our chances of success in the struggle for survival. If the human animal is better off not knowing certain things, natural selection will oblige by ensuring that the membrane of consciousness is only selectively permeable.

Human social groups simmer with complex, multiple, simultaneous interactions that would overwhelm a nonmodular mind. We could not possibly keep track of them without the benefit of demons specialized to monitor the tangled strands of our social relations. Self-deception is perhaps made possible by a system of unconscious filters that select what information is suitable for public consumption (conscious awareness) and what must be held back. The choice of what to hide and what to reveal is based on an unconscious assessment of what is most likely to be advantageous in the politics of social life. Now, this is all fine in theory, but what about the evidence? How does this present itself in our daily dealings with one another?

My account of how this happens, which will begin in chapter 5, is by far the most controversial message in this book. It addresses a phenomenon we all experience, but which, I will argue, few people recognize and understand. In effect, I will begin to peel back the protective covering that conceals the depth and extent of Machiavellian maneuvering in everyday social life, and in doing so, introduce the unconscious module that plays a central role in these dynamics.

5

Social Poker

Your face . . . is a book where men may read strange matters.
—Shakespeare, from *Macbeth*

Evolutionary theory implies that minds strive to maximize the reproductive success of their owners. Nature selects whatever works, whatever aids us in the struggle for survival, and self-deception works a lot of the time. As Trivers grimly reminds us, "The conventional view that natural selection favors nervous systems which produce ever more accurate images of the world must be a very naive view of mental evolution."[1] There is no arguing with the fact that it is seriously unsettling to entertain the thought that we spend our lives being duped by our own minds. The mirror of consciousness distracts us by projecting a flattering self-portrait, leaving us free to pursue self-serving social machinations unconsciously, in the dark, undisturbed by conscience and unburdened by guilt.

Playing Poker in the Dark

Humphrey's chess metaphor mentioned in chapter 3 depicts social interaction as a game, an open-ended sequence of tactical ploys between players. We play for high stakes, and oddly, social life is a game we play best when we do not know that we are playing it.

Competitive play demands that each player keep an eye on the other's moves. A tennis player cannot return a serve if she is not aware of where the ball is headed, nor can a boxer avoid a blow if he does not track his opponent's fists. The art of mutual scrutiny has an ancient lineage. A predator on the prowl must carefully observe, and try to second-guess, the evasive movements of its prey, which must in turn anticipate the tactics of its predator. Children, whose games prepare them for life in a dangerous world, endlessly rehearse these scenarios. The game of "red light, green light" or "statues" hones the freezing response that confounds the predatory gaze, while "hide-and-seek" drills children in the skills required for evading capture and getting "home" safely. Our minds are anticipatory engines, equipped by natural selection with intimations of the tasks required for survival. At that fateful moment when our species became its own main predator, these well-practiced cognitive routines swung into action in the social arena. Skills originally hammered out in our ancestors' dealings with predators and prey were put to use as methods for flourishing in the social equivalent of the African savanna.[2]

To be of use, the unconscious mind has to do more than just perceive. It must also be able to influence behavior. There must be some mechanism enabling it to use unconscious information to choose a course of action while at the same time keeping the conscious mind from getting wind of what is really going on. To do this, the unconscious mind must be agile, built for speed, capable of thinking several moves ahead in less time than it takes the lumbering conscious mind to take one laborious step forward. The effective deceiver must be able to track others' responses on a moment-to-moment basis, adjusting his or her tactics based on a steady stream of perceptual feedback. An artful wheeling-dealing species *must* have a knack for predicting, controlling, and understanding behavior. In order to do so, it must have an intuitive grasp of how to infer others' mental states and how these mental states work together to produce behavior. For this reason, we spend a good deal of our time trying to figure out the mental states of others—their beliefs, desires, goals, and fears—so as to manipulate their behavior in ways that serve our own interests.[3]

A savvy social operator needs to have an excellent grasp of human self-interests, because it's impossible to beguile others unless you understand what makes them tick. However, self-deception, which is also essential for competent social manipulation, pulls us in the *opposite* direction, leading us to disavow knowledge about human self-interest and encouraging a rather naive conception of human nature. So, there is a tension between the profound psychological understanding needed by the shrewd social player and the dumbing-down of social intelligence required by self-deception. How could nature have engineered the mind to make the most out of these conflicting forces? The obvious solution is to *split* the mind. It is all right for consciousness to be socially myopic if this helps social intelligence to operate smoothly behind the scenes. Our sheer conscious stupidity about one another is a perfect front for a wily Machiavellian intelligence.

If these considerations are on the mark, we must be far better at "reading" other people unconsciously than we are consciously. We must all be gifted, instinctive psychologists, whose conscious minds are left out of the loop. The chasm dividing unconscious astuteness from conscious naiveté is a consequence of the way that natural selection has sculpted our psyches to handle the pressures of a complex social life. It is distinctly and quintessentially human.

The game of life is starting to look a lot more like the fiercely psychological game of poker than the comparatively asocial game of chess. Deception and mind reading are essential components of poker. Remember the scene in the film *Maverick* in which James Garner defeats a femme fatale bluffing with a bad hand? When she asks him how he saw through her ruse, Garner remarks, "Breath holding. You hold your breath when you get excited." Maverick's mind reading feat is not unrealistic.

Accomplished poker players mobilize an army of techniques for misdirection, manipulation, and intimidation that have an uncanny resemblance to the methods used by other species to defend themselves against predators. They engage in crypsis—the famous "poker face"—to conceal their reactions and intentions or use incessant chatter—"talk a good game" as a diversionary tactic. "Berserkos" force others to make risky moves. Some, like "Crazy" Mike Caro, use the protean strategy of seemingly bizarre and random behavior.

> *Crazy Mike would discard cards at random (or pretend to), give away "bonus" chips to losing players, deliberately play with six cards (a dead hand), raise and bluff with no chance of winning, play a portable tape recorder indicating whether he would pass or raise, and challenge other players to "death matches" in which the loser must kill himself.[4]*

Caro was crazy like a fox. He used bizarre behavior to disorient the other players, to "jam" their mind-reading equipment.

Poker players make a great effort to read one another's "tells," the small involuntary information leaks that betray a person's true feelings or intentions, while working hard to suppress their own. "We are riddled with these tics and twitches," writes poker player/journalist Andy Bellin. "They are everywhere and in everything we do."[5] Top-notch play piles deception upon deception. If information leakage is involuntary and can be used by one's opponents to guide their play, why not *simulate* information leakage to deceive the other players? A master of the game may produce "false tells" by, for example, intentionally scratching his right cheek when bluffing, and then use this manufactured idiosyncrasy to get opponents to misread a winning hand as just another bluff. A person unable to comprehend the subtle, layered, social choreography that unfolds around the card table cannot function at the highest levels of the game where, according to poker lore, you "play the player."[6]

How do they do it? How is it possible to muster this kind of insight accurately and consistently enough to put it to use? Poker virtuosos do not draw upon some mysterious gift of intuitive insight. The price of their amazing abilities is hard work and study. "The ability to decode," writes psychiatrist Charles Ford, "is attained only by keeping exhaustive mental or written notes or by personal familiarity. In this respect, the considerable skills of professional poker players reflect the deception and detection abilities in everyday life."[7]

Most of us are embarrassingly inept at spotting liars. Even professionals make a poor show of it. In one study psychologists asked experienced law-enforcement officers, rookie cops, and college students to determine whether various individuals were lying or telling the truth. The results were disheartening. Not only were there *no* significant differences in the accuracy of the judgments of the three groups, but also all three guessed right at a frequency only minimally better than chance. Even the experts might as well have just flipped a coin! Another study used videotapes of simulated interrogations in which half of the actors were instructed to lie and the other half to tell the truth. When the experimenters asked police officers to distinguish the liars from the truth-tellers, they performed at the same level of accuracy as random guessing (they got it right only 49 percent of the time). These two studies are certainly open to criticism. We would not expect actors to be anxious about being found out, so they might fail to generate the involuntary information leakage (the poker player's "tells") that would be expected from real criminals with something to hide. However, more carefully designed studies have produced comparably abysmal results.

Although most people are very bad at noticing lies, a few, perhaps one in a thousand, are extraordinarily skillful. What is it that makes the difference? Ace lie detectives home in on facial expressions. As we have seen, one must pay attention to nonverbal indicators to understand others' true feelings. Every basic emotion has its own characteristic pattern of expression. Sometimes these seep through the deceptive façade and fleetingly slip across the liar's face. These "micros" may last for no more than a hundred twenty-fifth of a second. This is much too fast for an observer to consciously perceive (remember Libet's findings), but they are, nonetheless, picked up and analyzed by unconscious neural circuitry specially dedicated to this purpose. Another, more obvious, giveaway is the attempt to "squelch" the true feelings with rigid, unnatural facial expressions.[8]

The skill of intentionally unmasking deception does not come spontaneously, but it can be learned with education and practice.[9] The fact that lie detecting does not come naturally does not appear to sit comfortably with the idea of an evolutionary arms race between manipulation and protective mind reading. If deception was a selection pressure for the evolution of detection, which in turn was a selection pressure for the appearance of superior forms of deception, and so on, wouldn't we expect our species to be quite adept at detection? Does the evolutionary story not imply that lie detection should flow easily and naturally rather than being something that requires effort and training?

When dealing with any evolutionary adaptation, we need to

be mindful of the fact that it is fine-tuned to fit the environment in which it evolved. If the environment changes, the adaptation may no longer be able to work. Thankfully, our physical environment usually changes very slowly. Catastrophic events, such as the meteoric collision that exterminated the dinosaurs, are quite rare. The same cannot be said of our social environment. It is arguable that a major social upheaval long ago caused our neural mind-reading equipment to falter, and we have not yet recovered from that. The evolution of language was probably the single most potent event in our social evolution, and transformed the landscape of human relations beyond recognition. It may well have disturbed our hard-won ability to penetrate deception.

> With the advent of language in the human lineage, the possibilities for deception and self-deception were greatly enlarged. If language permits the communication of much more detailed and extensive information . . . then it both permits and encourages the communication of much more detailed and extensive misinformation. A portion of the brain devoted to verbal functions must become specialized for the maintenance of falsehood. This will require biased perceptions, biased memory, and biased logic; and these processes are ideally kept unconscious.[10]

Language ushered in a new phase in the timeless struggle between the forces of deception and detection. Loading the dice heavily in favor of the former, it enabled human beings to misrepresent reality much more effectively than had previously been possible. The sword of language cuts two ways: it is both enlightening and bewitching, an instrument for understanding and a snare for the unwary. The "gift of gab" allowed our ancestors to paint false portraits of the world, including their own motives, at absurdly little cost. It also enabled them to deceive themselves, or at least to deceive themselves more effectively than they had been able until then, and therefore to do a better job at deceiving others. Opinions vary widely about precisely when this epoch-making event occurred; and, given the scarcity of evidence, it is unlikely that we will ever be certain. Current estimates range from about 50,000 to about 500,000 years ago.[11]

This explains why acquiring the skills to spot deception involves more *unlearning* than learning. We have to develop the knack of listening to the music of human communication rather than concentrating on the lyrics. The would-be lie detector must discard their almost automatic privileging of the authority of speech over raw observation, of paying more attention to what people say than to what they do. In short, we are like people playing poker in the dark. We are in the dark because "part of the game of social competition involves concealing how it is played. . . ." from our own conscious minds.[12] As unconscious poker players, we can manipulate the others while remaining innocent of many of our own self-serving intentions. If accused, we can sincerely take offense and claim that it is all in the paranoid eye of the beholder. Self-deception about our own Machiavellian agenda also makes us relatively insensitive, on the conscious level anyway, to the selfishness of others. We sleep because waking up would spoil the game.

Psychoanalysis Redux

The Pinnochio problem was not lost on Sigmund Freud. "When I set myself the task of bringing to light what human beings keep hidden within them. . . ." he wrote in 1905, "I thought that the task was a harder one than it really is."

> *He that has eyes to see and ears to hear may convince himself that no*
> *mortal can keep a secret. If his lips are silent, he chatters with his*

fingertips; betrayal oozes out of him at every pore. And thus the task of making conscious the most hidden recesses of the mind is one which it is quite possible to accomplish.[13]

Freud held that the mind is opaque, riddled with self-deception, and mostly unconscious. He placed self-deception ("defense" or "repression") at the heart of his conception of human nature and believed that the recognition of unconscious mental processes would open exciting new vistas for psychological science. Freud also held that we must consider the human mind in its broader biological context, and proposed that the study of evolutionary biology should be part of the professional education of every psychoanalyst.[14]

These are eminently sensible ideas, which render the dramatic failure of psychoanalysis as a science and as a therapy even more striking. Although magnificent in conception, psychoanalysis never delivered the goods. Where did Freud and his followers go astray? The failure of Freudianism came down to three crippling shortcomings. First, the Freudians did not recognize the need for empirical research that could bring their confident conjectures before the tribunal of evidence. Without this, psychoanalysis could not distinguish between true and false claims, remaining at best a speculative system and at worst a collection of dramatic but baseless fantasies. Second, although Freud wanted his brainchild to be a complete interdisciplinary science of the mind and therefore drew on what was then cutting-edge work in biology, neuroscience, anthropology, and other fields, these sciences moved on and left psychoanalysis behind. What was once cutting-edge science soon became a museum of quaint intellectual anachronisms. Psychoanalysis ossified in isolation.[15] The third reason for the failure of psychoanalysis is perhaps the most interesting. Freud simply did not have the tools to realize his ambitious dreams. Neuroscience was barely on its feet. Cognitive science, with its sophisticated techniques for studying unconscious mental processes, did not even get started until three decades after Freud's death. The modern synthesis of genetics and evolutionary biology was not yet forged; and, Hamilton's vital discovery of inclusive fitness, which made the emergence of sociobiology and evolutionary psychology possible, was still decades in the future.

In a 1985 interview for *Omni* magazine, Bob Trivers remarked, "Probably the most ambitious thing I want to do is to send Sigmund Freud to his grave once and for all. . . ."

I think Freud failed to establish a scientific methodology and tradition that would generate useful information for subsequent generations of psychologists. It's one of the scandals of modern psychoanalysis that more than seventy years have gone by and we still have so little scientifically usable information on key assumptions of the psychoanalytic system. I'd like to help lay the foundation that Freud failed to lay.[16]

In trying to understand the dynamics of the unconscious Machiavellian mind, it might be a good idea to check if there is anything that can be salvaged from the Freudian wreck that may help us build a new and better vessel. Picking through the debris, I have found a few treasures. Before describing them, I want to address some concerns that may be simmering in the reader's mind.

It would be entirely understandable for the scientifically informed reader to begin to feel some misgivings about the direction that this book is taking. After all, I have not provided one iota of empirical support for my claim that human beings unconsciously monitor one another in Machiavellian social exchanges. Sure, it's implied by the theory of self-deception, but is that good enough? Now, even worse, I am preparing to draw inspiration from a source that everyone knows is scientifically bankrupt. Counsel for the defense must have very little going for his case if he intends to call such a disreputable witness to the stand!

This pessimistic stance is understandable, if not entirely justified. It is true that the last thing we need is yet another speculative, pseudoscientific theory of the unconscious mind. But there are some mitigating factors in this case. Investigating the Machiavellian mind involves confronting facts that our own conscious minds have been designed by natural selection to evade. Although there are empirical investigations that support the existence of a module devoted to social reasoning and the detection of cheaters, these studies are concerned with *conscious* cognition and leave unconscious mind reading untouched.[17] What about the extensive psychological literature on unconscious mental processing? Does this have any evidence to offer? Unfortunately not, and for good reason. We are looking for mental processes that are only "switched on" when people become engaged in personally meaningful and emotionally significant social interactions, interactions where there is something at stake. Looking for them in the contrived and antiseptic conditions of a psychological laboratory is therefore doomed to failure. Like it or not, tracking down this elusive beast forces us to put on our hiking boots and trek out to wild, chaotic, naturalistic settings where real, emotionally charged interactions are found. We must become field biologists of the mind.

Of course, it would be wonderful to have a large body of pertinent scientific research at our disposal. Since we do not, we must use the limited data available to us without neglecting any clues, however anecdotal and possibly wrong-headed they may be. Although psychoanalytic theory is largely devoid of scientific value, the fact remains that psychoanalysts have been sitting behind the couch for over a century now, patiently listening to the emotionally poignant discourse of the consulting room. Although it would be foolish to use psychoanalytic writings as *evidence* for unconscious social cognition, perhaps all this listening has turned up observations that will be useful to us.

The Machiavellian Module

A trawl through Freud's writings turns up a few gold nuggets among the rubble. Consider the following comments Freud penned in 1913: "I have had good reason for asserting that everyone possesses in his own unconscious an instrument with which he can interpret the utterances of the unconscious in other people."[18] He also remarked that:

> *Psycho-analysis has shown us that everyone possesses in his unconscious mental activity an apparatus which enables him to interpret other people's reactions, that is, to undo the distortions which other people have imposed upon the expression of their feelings.*[19]

We can spell out Freud's proposals more fully, translating them into a contemporary idiom, as follows:

1. The architecture of the human mind includes a module (Freud's "instrument" or "apparatus"), the function of which is to accurately infer ("reconstruct") other people's mental states.

2. This module operates unconsciously.

3. It is *domain-specific*. That is, it is attuned to the emotionally charged mental states concealed by deception ("the distortions . . . which other people have imposed upon the expression of their feelings").

4. The input of the system consists of the disguised expressions ("derivatives") of other people's unconscious mental states.

Freud's remarks point to the presence of a mental module specialized for interpersonal perception, which is able to reach behind the deceptive veneer that often covers our real attitudes toward one another. I call it the Machiavellian module.

It is frustrating that Freud does not explain how he reached his conclusions. If we knew what observations were at the bottom of these claims, they might provide some information about the *output* of the system, the way that this module affects human behavior, and therefore give us some idea of how to go about observing it in action.

A search of the wider psychoanalytic literature shows that nobody else picked up this line of investigation until the eccentric Hungarian psychoanalyst Sándor Ferenczi delivered a paper at a conference in Wiesbaden, Germany, in 1932.[20] The paper, which was not well received, asserted that individuals undergoing psychoanalytic treatment often "have an extremely refined feeling for the wishes, tendencies, moods, and dislikes of the analyst, even should these feelings remain totally unconscious to the analyst himself."

I do not know whether they can tell the difference by the sound of our voice, by the choice of our words, or in some other way. In any event they display a strange, almost clairvoyant knowledge of the thoughts and emotions of the analyst. In this situation it seems hardly possible to deceive the patient and if such deceit is attempted, it can only lead to bad consequences.[21]

If true, this implies that the Machiavellian module is exquisitely sensitive to both the conscious and unconscious mental states of other people, and is relatively impervious to interpersonal deception. What about the output question? We are given a hint in an obscure remark toward the end of the paper referring to a "strange, much veiled, yet critical manner of thinking and speaking," implying that the output of the Machiavellian module is verbal.[22]

The Wiesbaden paper was Ferenczi's swan song: he died of pernicious anemia later the same year. However, in the diary that Ferenczi kept during the final year of his life, there are two brief entries with a bearing on our question. Both of these involve Ferenczi's patients reporting recollections to him. Ferenczi interpreted these recollections as unconscious portrayals of the real implications of his (Ferenczi's) own behavior. The canonical psychoanalytic take on memory evocation has it that childhood memories and intrapsychic conflicts unconsciously *distort* conscious perceptions of present-day relationships, and that the influence of the unconscious is therefore fundamentally maladaptive. Ferenczi proposed the exact opposite: the memories were evoked precisely because they were a good analogy for the real, present-day social interaction between analyst and patient. They represented a disturbing aspect of reality all too well. In one case, a patient spoke at length about a childhood teacher "who was very nice . . . yet always maintained a pedantic attitude."[23] In another, a patient reported a dream about a "helpless struggle" to communicate a message to a man who did not understand her. In both cases, Ferenczi interpreted the patient's remarks as a reflection of their relationship with him: his pedantic attitude in the first, and his inability to understand in the second. Perhaps without realizing exactly what he had stumbled upon, Ferenczi developed the rudiments of the idea that we unconsciously respond adaptively and perceptively to current social situations, and that we convey these perceptions to others in disguise.

Memory and Mind Reading

Ferenczi's hypothesis raises some significant issues that repay careful consideration. Why it is that memories sometimes intrude unexpectedly into consciousness? The obvious explanation is that some environmental stimulus triggers them, in much the same way that a particular smell or piece of music has the power to transport one back to an earlier time. The kind of memory evocation that we are concerned with here is more opaque. When you hear a song on the radio that arouses memories of, say, your senior year in high school, the connection is immediately apparent. More often than not, though, memories are aroused with no obvious connection to a precipitating trigger, as I will shortly show with some examples. Ruling out the unscientific notion that these are simply random events, not caused by anything at all, it seems that the likely culprits are stimuli of which we are not aware. This explanation does not take us very far because it is much too imprecise. We need a reasonably clear idea of what kind of stimuli triggers these memories, and why they do so.

Ferenczi's examples suggest that the unconsciously discriminated meaning of an environmental stimulus is what primes at least some recollections. This invites a further question: Why is it that not all of our perceptions have the power to trigger memories in this way? If you consider the first example from Ferenczi's diary, there must have been a good deal more going on in the room than was represented by the memory of the pedantic teacher. Why is it that Ferenczi's patient did not respond unconsciously to the color of her analyst's shirt or to the vase of flowers on the table? Why is it that only certain perceptions operate as memory pumps? The answer is obvious. If *you* had been lying on Ferenczi's couch, wouldn't his condescending attitude have felt far more significant than the flower arrangement?

In any social encounter, we spontaneously give priority to concerns about conflicting interests, deception, and manipulation. When confronted with another person whose behavior matters to us in some way, we unconsciously monitor that person's actions and expressions and then, delving deep into the memory library, select images that resonate with the meaning of the interaction. Ferenczi called these memories the "historical dressing-up" of a "wholly contemporary" situation.[24] Although set in the near or distant past, they point to the unspoken dynamics unfolding in the emotionally charged present. By disguising our perceptions as memories, we are able to speak the unspeakable, to talk about the raw realities of social life while preserving the integrity of self-deception.

An African Orphan and a Chick Too Young to Fry

At this point, I think that it will be useful to leave the realms of theory and consider some examples that I think show the Machiavellian module in action.

My wife, Subrena, is twenty years younger than I am. She is a woman of color and I am a Caucasian male. For many people, the sight of a middle-aged white man married to a younger black woman activates a variety of familiar stereotypes, fantasies, and suspicions. Although never mentioned in polite conversation, they find indirect and unconscious expression. One such incident occurred shortly after we had moved from England to the United States. Subrena and I attended a faculty party and were introduced to a white male colleague who, after a few minutes of small talk ("How are you enjoying Maine?"), filled an uneasy silence by suddenly informing us that his cousin had recently adopted a child from Africa. He emphasized that his cousin's decision was unwise because "he's really too old for that sort of thing." Of course, it is possible that colleague's selection of precisely this conversational topic was purely coincidental, but the alternative explanation is almost too obvious. This man's story turned on the themes of age and race: a Caucasian man who was too old adopted a child from Africa. It does not require an enormous leap of the imagination to conclude that it had something to do with his reaction to my wife and me. It is also obvious why he expressed these sentiments unconsciously. It would, after all, do violence to the rules of civilized social discourse to blurt out, "You are too old and too white to be married to this young black woman." We have experienced this sort of thing many times. On another occasion, after a similar introduction at a party in London, a male colleague started talking to me about the African-American jazz musician Louis Jordan. He then began to sing an obscure piece from Jordan's oeuvre. It was a song entitled "That Chick's Too Young to Fry," the lyrics of which humorously advised a man dating an underage girl to break off the relationship until she was older.

When I tell these and other similar stories in the classroom or at public lectures, most people intuitively understand the subtext and do not find it particularly implausible. However, I often receive a very different response from psychologists, who say something like "Yes, very amusing; but it's not science." Of course, they have a point. It is true that anecdotes are not science, but a lot of science begins with anecdotes that point toward phenomena worth investigating. If this phenomenon is real, it suggests the existence of aspects of the human mind ignored by mainstream psychological science. This is certainly something worth being curious about.

There is no denying that these concepts will sound wildly implausible to many scientifically minded people. The problem lies less in the notion of an unconscious module for social cognition, which enjoys a good deal of support in the scientific literature, than it does in the idea that this module expresses itself through unconscious verbal communication. To many, this claim goes far beyond anything that science could possibly endorse. My next task, then, is to show you that the idea is less bizarre than it might at first appear. This will be the focus of chapter 6.

6

Hot Gossip

The universe is made of stories, not atoms.

—Muriel Rukeyser

We have seen how unconscious social communications can pack two distinct meanings in a single utterance: a surface (explicit, overt, conscious) meaning, and a coded (implicit, covert, unconscious) meaning. Coded meanings have to be interpreted in order to be understood. This poses a problem because the very idea of interpretation has acquired an unsavory reputation among members of the scientific community.

Interpretation has fallen upon hard times, and is nowadays barely taken seriously by anyone working outside the humanities. Gone are the days when priests scrutinized and decoded the entrails of sacrificed sheep to forecast the emperor's future. The practice of interpretation has a continuous unscientific pedigree extending from the attempts of the ancient augers right up to modern dream analysis. Because of this, anyone with a serious interest in interpretation runs the risk of being accused by scientific colleagues of having gone off the deep end, of doing literature rather than science, or of being a born-again Freudian.

I mentioned in chapter 4 that psychologists defined their discipline during much of its history as "the science of the prediction and control of behavior." Psychologists eventually came to accept that in order to predict and control behavior, one must first *understand* it, and that understanding human behavior is not even conceivable without resorting to the sort of "mentalistic" concepts despised by the behaviorist hard-liners. In fact, the very notion of behavior depends on concepts such as intentionality, action, and decision. (If you doubt this, try explaining why someone else raising your arm does not count as an example of your behavior, while raising it yourself does.) Part of what makes a bodily movement into a behavior is its meaning, a mentalistic concept if ever there was one. To top it off, philosophers have shown that it is impossible to understand one meaning in isolation from a whole network of others. Meaning is, so to speak, *spread out* over the mind. Philosophers call this the "holistic" character of meaning.

The holistic character of meaning makes investigating meanings scientifically a very daunting prospect. The standard equipment in the scientist's intellectual tool kit is not up to the task. Scientists work by segregating the phenomena that they want to study. For example, if a scientist is investigating the properties of a newly synthesized chemical, he or she attempts to obtain a pure sample of the substance and test the way that it reacts with other substances. It is crucial for the scientist to exclude all "confounding variables"—that is, extraneous factors that may influence the results of an experiment. It is impossible to investigate meanings in this fashion. Words simply cannot be isolated from one another, or wrenched out of the contexts in which they are used. Imagine three scenarios in which a man says to a woman "How about another date?" In the first one, he is holding an open box of sugared dates in her direction. In the second, he has just walked her home from a romantic evening out. In the third, he is her history professor and she has just said "1968" in response to the question "When did World War II begin?"

Ludwig Wittgenstein and a number of philosophers after him argued that "meanings" are not entities tacked on to language, and that we do not give words meaning by simply assigning them in piecemeal fashion. The meaning of a word—or a sentence—is its use: words mean what words do, and how they cooperate with all of the other words and sentences in a language to do their job. It is not possible to investigate the meaning of a word, say "seat," by compiling statistics on the circumstances in which the word is used. We do not utter the word "seat" only in the presence of objects to sit on; we often use the word in their absence (you have been on your feet all day and would "give anything for a seat"). Used as a noun "seat" denotes a very diverse collection of objects including chairs, sofas, bicycle seats, and even crates, boulders, logs, and so on. But even this is far from the whole story. What about "escaping by the seat of your pants," or "the county seat," or "the seat of the soul"? To comprehend language we have to feel our way into it.[1]

Meaning and interpretation are inescapable, and any adequate theory of how the mind works must eventually come to grips with them. Interpretation is quite a commonplace activity. We have to interpret the meaning of what people say and do virtually every day of our lives. We also have to interpret meaning every time we take a crack at a crossword puzzle, read poetry, enjoy puns, or engage in innuendo (as my grandfather used to say, "Love goes out the door when money flies innuendo.").

Any scientific approach to interpretation must find a way to surmount a fundamental methodological problem. Would-be interpreters have to find some way to make sure that they are not just seeing faces in the clouds, attributing meaning willy-nilly in a manner that is neither testable by any conceivable empirical procedure nor constrained by evidence. It is easy to mistakenly read unconscious meanings *into* things, so unless there is some way to establish whether or not messages *really* mean what one takes them to mean, interpretations, however clever or compelling, cannot reasonably be treated as anything more than either inspired conjectures or ingenious fantasies.

In the following chapters, I will sketch out an approach to the interpretation of unconscious meaning that is, if not watertight, at least constrained by evidence. Future research may be able to provide a more definitive approach to the problem of interpretation. What is most important right now is the relatively modest goal of providing grounds for taking interpretation, and unconscious meaning, seriously. It would be foolish to ignore phenomena that stare us right in the face just because they do not easily yield to the standard methods of experimental science.

A Cure for the Common Code[2]

One way to approach the subject of unconsciously coded communication is the more intuitively accessible route of conscious encoding. We are all familiar with double entendre. "Is that a gun in your pocket?" purred Mae West, "Or are you just glad to see me?" The technique of double entendre enabled West to speak of subjects that were at the time unmentionable in polite society, to hilarious effect. This technique is not restricted to the conscious domain. Sometimes we come out with double entendres unconsciously, such as the man at a party who decides to step outside for a moment and, encountering an attractive woman exposing an expanse of cleavage, mumbles, "Pardon me, I need to get a breast of flesh air." Although inadvertent, and literally nonsensical, the meaning of this slip of the tongue will not be lost on anyone who understands its context.

Sometimes we encode information for purely practical reasons. Codes such as the Morse code or the "One if by land, two if by sea" that Longfellow put into the mouth of Paul Revere can aid the transmission of a message. Other codes, such as shorthand, assist transcription. Coding can also enhance the efficiency of storage: the binary code used by my PC allows me to house a bulky manuscript inside a frail floppy disk.

The purpose of some codes is to conceal information from third parties. Military codes, such as the infamous "Enigma" code used by the Germans during WWII, are well-known examples of this type. A more mundane example is parents' practice of spelling out words that they do not want their young children to understand ("the c-o-o-k-i-e-s are in the c-a-r"). Sometimes coded messages are attempts to get a third party to *falsely* believe that they understand a message. The spirituals sung by African-American slaves during the nineteenth century used religious lyrics to conceal a subtext about escape and resistance that was impenetrable to the ears of the slave-owners and bounty hunters.

Military coding typically makes use of an arbitrary, digital system of symbols (for instance, representing letters of the alphabet by numbers). In contrast to this, Mae West's double entendres and the lyrics of the slaves' songs expressed hidden messages by using analogy. Analogical coding differs from digital coding in that it represents by resemblance: the symbol actually looks like the thing that it symbolizes. The word C-A-T (a digital code) looks nothing like the furry animal that purrs and rubs against your ankles. A stylized drawing of a cat, on the other hand, actually resembles the beast, albeit on a rather abstract level. Mae West's "Is that a gun in your pocket . . ." works because the protuberance caused by a revolver in a man's trouser pocket resembles the visual effect produced by an erection. A song lyric about crossing the river Jordan to the Promised

Land delivers its subversive punch because of the resemblance between this and the act of crossing the Ohio River to the free states.

The examples of unconscious communication in the previous chapter, the stories of the African orphan and the chick too young to fry, are analogical. In each case, a social situation evoked a memory that analogically expressed part of the Machiavellian meaning of the immediate situation. These memories were not the kind that we draw on when we recall, say, that Napoleon died on St. Helena ("semantic" memory), and are obviously different from "procedural" memories such as remembering how to ride a bike. Unconscious communication takes the form of "episodic" memories: memories of particular happenings. Can you recall what you did right after rolling out of bed this morning? How about the plot of the most recent movie you saw? Can you recall what you did last New Year's Eve? These are all examples of episodic memories.

Episodic memories tell a tale. So, from here on in, I will drop the term "episodic memory" in favor of the more flexible, evocative, and intuitively appealing words "story" and "narrative." Stories depict real or imaginary scenarios that unfold in time. They are about things that happened, are happening, or are going to happen. They are told, sung, danced, or enacted. Stories concern "concrete individuals in concrete situations": news reports rather than editorials.[3] You know that you are hearing a story when you can mentally project the sequence of events that it describes on the movie screen of your imagination. Although nonnarrative discourse may be informative, only narratives can be *gripping*. Stories seize the mind.

Of course, not every story is an unconscious message (as Freud reportedly said, "Sometimes a cigar is just a cigar"). People tell stories for all sorts of reasons, some of which are fully conscious and patently obvious. If we are to press this line of investigation further, it is crucial to be able to distinguish between consciously and unconsciously motivated stories. What follows are some guidelines that, although rough and ready, will serve us adequately for now.

First, as outputs from the Machiavellian module, unconsciously coded messages are likely only in certain social contexts. People deliver unconscious messages in circumstances where it is important for them to determine the concealed intentions of other people, situations in which manipulation, misdirection, and self-deception are likely. We should expect to hear unconscious messages in situations involving *covert conflicts of interest* between the speaker and some other person(s), and in which it would be *disadvantageous to speak openly* about these conflicts of interest.

Second, coded narratives characteristically seem irrelevant to the conversational setting or discontinuous with the conversational flow, coming from "out of the blue." Consider the example of my colleague's reaction to my wife and me. His story seemed weirdly unrelated to anything going on in the immediate environment. Sometimes a speaker will leave an otherwise coherent conversation to veer off on an analogical riff—a seemingly irrelevant or gratuitous narrative tangent. Somewhat less tangibly, unconsciously coded communications often seem to have a sense of urgency. The subjective experience is something like "I don't know why I want to tell you this but I feel strongly that I want to tell you *right now.*"

A Winter's Tale

One exceptionally bitter winter morning, just before class was to begin, three undergraduate students entered the classroom and took their seats. The class was very small under the best of conditions, consisting of only seven students, but only Sara,

Amy, and Michelle put in an appearance that day. I decided to wait a while before kicking off the lecture, in case there were other students en route. As the three students began to chat, the following dialogue unfolded.

Michelle: I wonder where everybody is?

Amy: I heard a horrible story on the news, but I can't remember what it was.

Michelle: There was this guy who drove up into the mountains with his three-year-old child. He went out hunting and left the kid all by himself in the truck. When he came back his son was frozen to death. He just went off to enjoy himself, and when he came back his son was dead.

Sara: That's horrible.

Amy: Wasn't Tom Allen supposed to be here today? What time is he going to come? [This referred to a congressional representative who was scheduled to visit the university that day.]

Amy [turning to Sara]: I want to do a course with Professor H. next semester. He's so cute. He gets really excited when he teaches.

Sara: You can't. He's away on sabbatical. He won't be back till next year. Once in a while I see him in the supermarket.

What is going on here? On the face of it, not much more than idle chatter intended to fill the time before class starts, but quite a different story is revealed if we treat it as an unconsciously coded conversation. Michelle, Amy, and Sara found themselves unexpectedly confronted with a significant social event: half the class was absent. This event was biologically significant because the absent students defaulted on an implicit agreement to attend each session. I am sure that all three of them would have preferred to spend an extra hour under the blankets on that icy morning, but they dragged themselves out of bed and trudged through the snow to get to class. In other words, the students who chose to come paid a *cost* that the absent students did not have to pay (or, to turn it around the other way, the absent students enjoyed a special benefit). There were also much more subtle social forces at play, as none of the three women knew how the others assessed the situation. If one were to denounce the absent students, she might alienate her two peers, who in turn might report her attitude to the absent students, the most likely outcome of which would be social ostracism by the group.

Michelle was the first to speak up. She was a person who often dominated class conversation, and who cared intensely about the quality of other students' commitment to the course. In fact, she consciously mentioned the absent students at the very beginning of the conversation, but did so in a way that was entirely neutral, expressing only mild surprise. It was only after a prompt from Amy that she continued and seemingly changing the subject, recounted a sordid tale of callousness, selfishness, and the abnegation of responsibility. On the surface, this report had no clear connection with anything that was of concern to these three young women. Given the context, this should alert us to the possibility of an unconscious message. Listening to it as such, the coded meaning practically leaps out. The man in the story appears to stand for the absent students and his abandoned child stands for the three students who turned up for class. It sounds as if Michelle was accusing the absent students of selfishly ignoring their obligation to come to class. Perhaps this unconscious characterization of the situation was a way for Michelle to probe the attitudes of the other two. If so, Amy's brief reference to the congressional representative who was *supposed to be there* showed that they were on the same page.

Like the politician who had not yet arrived, all of the students had made a commitment to be present, although more than half of them had defaulted on it. The fact that Sara next picked up the ball and mentioned that Professor H. is *away on sabbatical* reinforces this interpretation. By the end of the exchange, all three students had unconsciously formed a coalition by establishing that they were all concerned about the absence of the other students. Although Michelle was the only one to provide a detailed and strongly negative assessment, neither of the others dissented from it.

Rather than overtly denouncing the absent individuals, Michelle's Machiavellian module selected a story that painted a picture of the delinquent group members as selfish and neglectful of the needs of the other students. The story also included the crucial element of the wintry weather, which the three young women had endured to get to class, and which the absent students had managed to avoid. My next point risks straining even the sympathetic reader's credulity to the limit. Notice that number three, which corresponded to the number of students present, also appeared in Michelle's narrative. We will return to, and consider in greater detail, the possibility of unconscious numerical references in chapter 8.

There is another feature of Michelle's story that is typical of unconscious communications: her reaction to the missing students seemed wildly exaggerated. It would be natural and understandable for Michelle to describe the missing students as somewhat irresponsible and self-centered, but Michelle unconsciously made them out to be cold-hearted criminals. Similarly, the students who turned up for class were young adults who were mildly inconvenienced by the situation, but they are unconsciously depicted by the heart-rending image of a cruelly neglected, helpless child. This sort of speaker-biased hyperbole is exactly what evolutionary theory would lead us to expect.

After all, it is the job of the Machiavellian module to help us pursue our interests and avoid being exploited in the treacherous game of life; it has no obligation to be "objective" in doing so. We are dealing with hot cognition rather than cool, detached reasoning.[4] Our unconscious assessments are uncompromising and unashamedly self-interested. The eighteenth-century philosopher David Hume captured the essence of this principle in his famous remark that "Tis not contrary to reason to prefer the destruction of the whole world to the scratching of my finger."[5] Mel Brooks made a similar point 300 years later: "Tragedy is when I cut my finger. Comedy is when *you* walk into an open sewer and die."

The conversation between Amy, Sara, and Michelle contained a remarkably elegant interweaving of pertinent themes. The choice of imagery was precise and highly economical. It was also automatic, slipping off the tongue without conscious deliberation or effort so naturally as not to arouse even a tremor of suspicion in the conscious minds of either the speaker or the listeners.[6] From their point of view, they were just having a chat about nothing in particular. When I pointed out to them what had gone on, all three burst into surprised laughter.

Again, it is possible that this is all coincidence? What guarantee is there that I am not just reading these meanings *into* the conversation? What makes this conception of unconscious communication anything more than yet another wacky, unscientific account of human behavior? Although my interpretation may sound plausible or even compelling, the fact remains that reporting a conversation between three students does not prove anything. It is very easy for an author to select an example that dramatically confirms a pet theory. Of course, I chose every vignette in this book precisely because each compellingly illustrates what I purport to be true. How could it be otherwise? There would be no point in offering them if I did not think that the phenomenon is genuine and my explanation of it largely correct. I am not trying to *prove* anything by these anecdotes. I am trying to get you, the reader, to entertain the idea that there might be something in this.

There must be *some* way to explain the thematic contents of seemingly idle talk. At the very least, my hypothesis has the advantage of being rooted in evolutionary biology and is consistent with much of what science tells us about architecture of the human mind. It also offers an integrative framework that casts new light on other investigators' results (more about this in chapters 7 and 8). These considerations fall far short of entailing its truth, but they bolster the case for taking it seriously. I will argue in the chapters to come that unconscious communicative phenomena are lawful and systematic, and can be tested using reasonably conventional methods, as well as less formal ones in everyday life. All of this will become clearer as we proceed. For now, let us return to the main topic of the present chapter.

A Brief (Pre)History of Gossip

Three women gossiping can present more than meets the ear. To make headway understanding unconscious messages, we need to have a close look at the origin and dynamics of gossip, broadly conceived of as the exchange of socially valuable and therefore "sensitive" information about members of one's community.

In spite of its magnetic appeal, gossip has had very bad press throughout the ages. Condemned in both the Old and New Testaments, as well as other religious teachings, it has sometimes even been made a punishable offense. Before the 1700s, gossip was punishable by the dunking stool, stocks, scold's bridle, and a nasty device known as the "branks," an iron mask with a spike that projected into the wearer's mouth. These European methods were mild compared to the treatment meted out by the Ashanti, who cut off gossipers' lips.

The eighteenth-century writer James Forrester noted, we regard "a malevolent Babbler with a worse Eye than a common Thief."[7] From the cautionary tales of the Middle Ages onward, works within the Western literary canon condemn gossip. *School for Scandal, Othello, Vanity Fair, Middlemarch,* and *Emma* are just a few well-known examples. Although these works treat gossip as a vice or a failing, they also sometimes grudgingly acknowledge its naturalness. As David Garrick sharply pointed out in his prologue to Sheridan's *School for Scandal*:

> *A School for Scandal! Tell me, I beseech you,*
> *Needs there a school—this modish art to teach you?*
> *No need of lessons now;—The knowing think—*
> *We might as well be taught to eat and drink. . . .[8]*

Garrick was right. Research shows that we spend 80 to 90 percent of our conversation time talking about other people, a full two thirds of which concerns people in our immediate social network.[9] A good deal of the remainder is taken up with talk about public figures, such as politicians, motion picture and television characters, celebrities, and, of course, ourselves. The wide appeal of television programs such as *Oprah* and tabloids such as the *National Enquirer* speaks to our voracious craving to get the "real scoop." In fact, there are entire disciplines, such as anthropology, counseling psychology, and cultural studies (not to mention history) that might be understood at least in part as academically regimented versions of the social grapevine. Our appetite for producing, consuming, and transmitting stories about one another appears to be of great antiquity, and is perhaps as old as speech itself. Investigations by Merlin Donald, an expert on the evolution of human cognition, suggest that:

Language, in a pre-literate society . . . is basically for telling stories. Language is used to exchange information about the daily activities of members of the group, to recount past events, and to some extent to arrive at collective decisions. Narrative is so fundamental that it appears to have been fully developed, at least in the pattern of daily use, in the Upper Paleolithic. A gathering of modern postindustrial Westerners around the family table, exchanging anecdotes and accounts of recent events, does not look much different from a similar gathering in a Stone Age setting. Talk flows freely, almost entirely in the narrative mode. Stories are told and disputed; and a collective version of recent events is gradually hammered out as the meal progresses. The narrative mode is basic, perhaps the *basic product of language.*[10]

According to British psychologist Robin Dunbar, whom we met in chapter 3, our ancestors learned to speak because they needed to gossip. Although this idea may at first sound ludicrous, it has some impressive research in its favor. In order to thrive in complex social groups of any size, group members need to network, to forge alliances. An accomplished social player knows how to win friends and neutralize enemies, to be in with the in-groups and stay out of the out-groups. These principles are not exclusive to our species. Nonhuman primates establish and maintain alliances by an activity known as "grooming." Primates groom one another by gently brushing back their partner's fur with one hand and using the other hand, lips, or teeth to extract foreign matter or pieces of dried skin. This intimate procedure can go on for seconds or for hours and is literally intoxicating, triggering the release of cascades of endorphins in the recipient's brain. Primates do not groom each other promiscuously or haphazardly; they form relatively stable and exclusive grooming cliques. The benefits of belonging to a grooming clique are far greater than the direct payoff of a daily massage. When a primate encounters trouble, members of its grooming clique are the ones most likely to lend a helping hand.[11]

There is a strong relationship between the size of a primate group and the amount of time that its members spend grooming one another: the bigger the group, the more grooming time. The social math is simple: the larger the group, the more networking required for social success. As we have seen in chapter 4, the complexity of a group increases exponentially with each additional member. This puts a cap on the maximum size of groups that rely on grooming to maintain social coherence. Beyond a certain point, too much time would have to be spent on grooming, and there would not be enough hours left in the day for vital activities, such as finding food.

There is also a robust correlation between the size of a primate's neocortex, the thinking part of its brain, and the group size normal for its species: the bigger the social group, the beefier the cortex. This provides a remarkable corroboration of the Machiavellian intelligence hypothesis. Complicated social systems require high intelligence, and therefore select for large cerebral cortexes. Primates living in smaller groups can afford to have smaller brains. Note that having a large brain is biologically costly: the brain is an expensive organ to run, and large ones consume many precious calories (the human brain accounts for a whopping 20 percent of our energy expenditure). For social primates living in comparatively large communities, the advantages conferred by social intelligence more than make up for the drain on resources. However, the increased nutritional demands made by a large brain may also limit group size in populations living in areas where calorie-rich foods are scarce.

It is likely that our ancestors groomed one another and formed intimate alliances with their grooming partners just like chimpanzees do today. Working backward from estimates of brain size based on skeletal remains, and applying the formula expressing the relationship between brain size and group size, Dunbar calculated that archaic *Homo sapiens* probably lived in communities consisting of about 150 individuals. Using the correlation between group size and grooming time, a group of 150 individuals would require its members to dedicate approximately 40 percent of their waking time—more than 9 hours a day—to grooming each other.[12] This would obviously be wildly impractical. No primate could survive in the real world if social activity monopolized his or her time to this extravagant degree.

What biological innovation made it possible for our ancestors to live in relatively large groups while maintaining a sufficient level of social cohesion? The most likely candidate is the evolution of language. Language is a far more efficient method for servicing alliances than old-fashioned grooming. It allows us to massage others' egos rather than just their bodies; several individuals can be simultaneously serviced in contrast to the strictly one-to-one grooming scenario; and, it facilitates vastly richer exchanges of personal information. Now, what kind of talk can we imagine our Pleistocene predecessors used to grease the wheels of their social networks? The first conversations must have been social talk. The primal human idiom was probably gossip.

It is likely that this momentous development first took root in the *female* brain, and only subsequently spread to males. Like present-day working mothers, Stone Age females had much greater demands on their time and energy than their male counterparts. Like females of many other species, human females pay disproportionately high reproductive costs. The total biological cost to the human male for the privilege of reproducing amounts to a couple of teaspoons of semen, which can be regenerated in a matter of hours. In contrast, women get a raw deal. They endure nine long months of pregnancy, which not only compromise their ability to perform everyday activities, but also expose them to health risks and create increased nutritional demands. This is followed by a risky and painful delivery, the physiological stress of lactation (nursing an infant takes time and energy), which continues for three years or so in hunter-gatherer societies, and responsibility for the care of a dependent infant. The additional energy demands alone are astounding. An average pregnancy costs a woman approximately 70,000 extra calories. This is not such a problem in the modern developed world, but during the Pleistocene (as in areas where poverty is rife today), it meant scrounging for extra nourishment. Compared to males, females had little leisure time and could not afford to devote long hours to grooming. By abandoning grooming for gossip, ancestral females found a way to reconcile the need for social networking with all of the other demands on their time and energy.[13] Language made it possible for a woman to multi-task; she could cradle her baby with her left arm and pick fruit with her right, while simultaneously exchanging gossip with a bevy of friends.

There is another reason to suppose that women were the first language-users. The exorbitant price that women pay for reproduction has influenced their sexual preferences. Research shows that women are attracted to males who are able and willing to provide resources to offset their reproductive costs.[14] Of course, this preference has also sharpened the deceptive talents of men, who may be more than willing to swear undying love just as long as it gets the object of their lust into bed. Words, after all, are cheap, and women are rightly inclined to treat declarations of love with a liberal dose of healthy skepticism. The capacity for gossip would have been a great boon to prehistoric women trying to assess the reliability of potential mates. Females could pool information about which men were trustworthy and which were rogues, tag them with positive or negative reputations, and use this information to guide mating decisions.[15] In fact, this vetting routine still goes on today. Modern women discuss with one another the details of their encounters with potential mates for this very reason.[16]

Gossip as Social Poker

We can now leave the origins of gossip behind and concentrate on what gossiping does and what adaptive problems it creates. This winding path will eventually lead us to some of the core dynamics of unconscious communication.

Gossip circulates vital information about whom to trust, and whom not to trust, transforming conduct into reputation by passing information through many mouths and ears.[17] In relatively small communities where it is impossible for an individual to blend into an impersonal and anonymous crowd, reputation is literally a matter of life and death. Those individuals known for their honesty and generosity are valuable coalition-members. Cooperators are popular and become resources for which other members of a community compete. Similarly, persons known to be treacherous, selfish, or unreliable elicit suspicion and may be ostracized or expelled from the group, a fate that in Stone Age conditions could easily have been fatal. Gossip establishes a person's status as a good or bad coalition-partner. The Yiddish writer Sholom Aleichem succinctly expressed the principle in his well-known epigram, "Gossip is nature's telephone."

No one with an understanding of Machiavellian intelligence can give unqualified credence to this benign conception of gossip as a clearinghouse for character references. If people with good reputations are a resource for whom others compete, this leads to all the dirty tricks that people use against one another when they are competing for something of value. One such move is to "poison the well," to destroy the perceived value of the resource. When the resource is a person's reputation, some individuals will spread malicious gossip to destroy or damage it (Roland Barthes described this as "murder by language").[18] Gossip can deceive rather than instruct and manipulate rather than inform. It becomes a form of informational aggression between individuals, and informational warfare between cliques.[19]

It plays with reputations, circulating truths and half-truths and falsehoods about the activities, sometimes about the motives and feelings, of others. Often it serves serious . . . purposes for the gossipers, whose manipulations of reputation can further political or social ambitions by damaging competitors or enemies, gratify envy and rage by diminishing another, generate an immediately gratifying sense of power, although the talkers acknowledge no such intent. Supplying a powerful weapon in the politics of large groups and small, gossip can effect incalculable harm.[20]

Unlike action, which requires commitment, gossip is a cheap signal that is easy to use self-servingly in circumstances where the interests of speaker and listeners diverge.[21] It displays what linguists call "displaced reference": that is, it refers to people and situations lying outside the immediate conversational setting. Information acquired from gossip is therefore difficult or impossible for the recipient to evaluate.[22] We have to take gossip either largely on trust, or not at all. Another major problem flows from the reciprocity of gossip. Gossip involves *exchanging* confidences, and these ostensibly reciprocal exchanges provide rich opportunities for cheating. It is always tempting to short-change another person, particularly if it is unlikely that they will ever discover the swindle. In the case of gossip, this means exchanging false or misleading information for information that is accurate. How can we know if the scandalous tales entrusted to us are true or false? Displaced reference makes it possible to trade valuable information for news that turns out to be false, misleading, or worthless.

The issue of privacy raises another vexing dilemma. Primate grooming is a public activity. It happens out in the open; so, in ape society it is obvious who is allied with whom. In stark contrast, gossip is essentially private talk, concealed from others' ears. Prefacing a juicy morsel of social information with phrases like "Just between you and me . . ." draws an invisible boundary around gossiper and gossipee, subversively setting them off from the rest of the community. "Gossip," writes Patricia Meyer Spacks, professor of English literature at Yale University, "creates its own territory."[23] This territory is only temporary: it may last only as long as the conversation, but each gossiper is a node in a wider informational network (to paraphrase John Donne, we might say that no pair of gossips is an island). The very person in whom you confide may soon be revealing something about you to a third party, perhaps the very person about whom you were just speaking. First Elaine tells Ann a titillating story about Paul, and then Ann, her tongue burning with this news, telephones Paul to inquire "Do you know what Elaine is saying about you?" Meta-gossip (gossip about gossip) adds a whole new level of complexity to the intricate currents of group information flow.

Once our ancestors learned to gossip, they could form secret alliances, deceive each other far more effectively about where they stood in relation to other community members, and stab each other in the back. Keeping track of multiple relationships that are out in plain view is arduous enough. The imposing challenge of monitoring multilayered clandestine relationships, of working out who is in cahoots with whom, of attempting to assess the reliability of secondhand information, must have taxed the prehistoric mind to the utmost. The advent of gossip foregrounded a whole new set of questions about social relations. Can this person be trusted as a confidante? Should I believe what they tell me? Are they manipulating me to further their own ends? Gossip *vastly* enlarged the Machiavellian dimension of social life and became a potent selection-pressure for the evolution of more and more penetrating forms of social cognition. The privacy of gossip, and the potential unreliability of the information that it circulates, introduced a terrifying note of *ambiguity* into human social life that was not present in the world of prelinguistic primates. These concerns require the potential gossiper to be cautious: only a fool would confide in another without carefully assessing the other's reliability. However, there is a catch-22: gossip demands give-and-take. As the ancient alchemists wisely wrote, "In order to make gold, you must first have gold": in order to receive information about others, one must first have information to exchange. *Obvious* reserve is counterproductive: it makes you seem suspicious and isolates you from the networks essential for social survival. One must appear to be open and trusting to encourage the flow of information, while at the same time carefully observing others for signs giving away their covert agendas. This is social poker with a vengeance. Just like virtuoso card players, conversational partners must rely on small, subtle clues to genuine motives that seep through cracks in unconsciously calculated displays of sincerity.

Cryptic Conversations

Let's now pause a moment for a reality check. Does this evolutionary account of social storytelling fit the facts, as far as we can discern them? Well, yes . . . and no. It is certainly true that gossip often circulates useful information about people in one's community, and that conspiratorial whispering and character assassination behind closed doors are all too prevalent; but, it is equally true that a great deal, perhaps most, of our chat about third parties neither makes nor breaks reputations, nor provides any clear benefit in the game of life.

Two friends sit down for a cup of coffee and one says to the other, "Did you watch so-and-so on TV last night?" and proceeds to describe the ins and outs of the episode. Now, no matter how entertaining it is, information about the plot of a situation comedy does not provide the listener with any information about fellow community members. It does not help the gossipers make mating decisions, decide whom to lend a tool to, or make any of the interpersonal judgments that the standard evolutionary take on gossip purports. Why spend one's time talking about a television program?

Evolutionary psychologist Jeremy Barkow suggests that the mental module evolved for gossip does not discriminate between television characters and acquaintances.[24] There was no television during the Pleistocene, and following the adventures of a character on a weekly basis is not unlike snooping on an acquaintance. The same template might be used to explain why we gossip about the lives of celebrities. However, this does not suffice for many other cases of seemingly trivial gossip. Why should an acquaintance suddenly regale you with an account of a chance meeting with a friend of her aunt Augusta? Why is it that this very morning three of my students, waiting for class to start, got into an animated discussion about filling a swimming pool with pudding, before drifting almost imperceptibly into a detailed discourse about *The Simpsons*? The prevalence of inconsequential chat does not sit well with the concept of information mongering that evolutionary psychologists seem to have in mind when they speak of the adaptive functions of gossip.

We need a principled biological explanation detailing *why* such interactions occur so frequently. More specifically, we need an account of why, at any given time, conversational partners select a particular, apparently trivial or irrelevant, topic to discuss. These questions may sound bizarre and it can be tempting to dismiss them, but it is not satisfactory to brush them off with responses such as "because it came to mind" or "because the speaker chose it." *Why* did the story come to mind? *Why* did the speaker choose it? In other words, *what caused the speaker to unconsciously select just this story, and no other, at precisely this moment, in exactly this context?*

The coded communication hypothesis gives us a possible explanation for at least some and perhaps many cases of seemingly inconsequential gossiping. We select particular stories at particular moments and in particular contexts because they portray the interpersonal dance unfolding right there and then on the interactional stage. Such stories encode information to shield our conscious minds from the complex Machiavellian dynamics humming away just beneath the surface of our encounters with others.

But Why Is It *Unconscious*?

Sitting on a bus crawling through heavy traffic in central London one muggy summer day, my wife and I found ourselves seated behind a young mother holding a fretful baby. The infant was hot, and the jerky movements of the bus and the still, stale air exacerbated her bad mood. The mother was perceptably irritated by the sluggish progress of the crowded bus: she was at wit's end and seemed unable to calm her infant. Predictably, the more stressed and irritable she became the more vehemently the baby cried. The outcome was an angry mother and a very distressed baby. We struggled unsuccessfully to "tune out" this little drama and carried on with a lighthearted conversation about various friends and relations, but we could barely hear ourselves think, much less talk. Subrena suddenly found herself launching into a vehement critique of a friend's style of child care. "I'm sorry, David," she said, "but I would never let Janet look after *my* child!" She went on to describe Janet's lack of patience with her daughter, how she would become enraged and inappropriately punish the girl in the mistaken belief that such "strictness" was good for her. At this point, the mother changed seats, moving farther away from us, and became calmer and more attentive. The baby stopped crying. It was at this point that Subrena abruptly ended her diatribe about Janet's dysfunctional mothering techniques.

Subrena's message seemed so transparent that I assumed that she had intended it quite consciously. When an opportunity arose, I asked, and she reacted with astonishment. She had been completely unaware of any connection between her story about Janet and the drama that had unfolded on the bus a few moments before. An obvious question to ask is why did Subrena do this *unconsciously*? Why did she deceive herself about the meaning and motivation of her story? Would it not have been more advantageous for her to respond consciously?

Just think of what would have happened if Subrena had directly accused the woman of handling the baby ineptly. The young mother would probably have taken offense and the conflict between mother and baby would have almost certainly escalated into something both more volatile and more costly. In concealing the meaning of her story from herself, Subrena did not have to struggle with issues about overstepping social boundaries. Furthermore, she could sincerely deny that she was trying to manipulate the young mother, who would not feel that another woman was challenging her maternal prowess. By communicating her message unconsciously, Subrena flew under the radar of both the mother's consciousness and, perhaps more importantly, her own. We will never know whether or not the woman on the bus consciously took my wife's remark as a hint, but based on many similar experiences, I very much doubt it. Unless the sequence of events was purely coincidental, the delivery of an unconscious message seemed to be a remarkably effective way of securing the desired outcome.

Mother Nature has seen to it that the conscious mind is relatively blind to the nuances of social behavior. It is easy to understand why this turned out the way that it did. If human beings had the knack of consciously making penetrating inferences about each other's motives and strategies, our insights would come at a high price. Self-deception would become much more difficult, and this would rob us of its vital benefits. To understand why, consider a physiological analogy. It is impossible for a person to damage his or her eyes in such a way as to make them unable to see only certain kinds of objects. There is no such thing as porcupine-blindness or the selective inability to see teacups. If one is blind, one loses a *whole dimension* of experience. The same principle applies to the social "blindness" of the conscious mind, which provides us with relatively impoverished portrayals both of our own actions and motives and those of others. All social inferences flow from a common set of assumptions, an informal folk-psychological theory of human nature. If the theory is biased, it will deliver faulty appraisals of *everyone*: not only of oneself, but also of other people. Commonsense assumptions include gems of sagacity such as the notion that self-deception is abnormal, that good people do not lie, that so-called normal people are not motivated by self-interest, and that politicians aspire to serve the public. Such homilies cannot serve as a basis for sound social reasoning, but they are terrific gimmicks for Machiavellian manipulation. The knife of self-deception cuts two ways: you cannot maintain a highly distorted conception of yourself side by side with a true estimate of others.

Take a walk with me through a dystopian world where conscious minds routinely deliver astute insights into the wellsprings of human behavior. In this imaginary but logically possible world, the people whom you encounter each day can read your motives as easily and accurately as they read the newspaper. Take a moment to ponder seriously what life would be like if we were all psychologically naked to one another's gaze. Social life would quickly unravel. Furthermore, given the deceptive character of human nature, this setup would be extremely unstable. Individuals who, through some genetic quirk, were able to conceal successfully their feelings and motives from others would have a competitive advantage, and the inexorable tide of reproductive success would eventually disperse the "deception gene" through the entire population. This, of course, would be the opening salvo in an evolutionary arms race. As the aptitude for deceit swept through the population, it would favor the evolution of more efficient methods of detection, initiating an arms race that would spiral through evolutionary time until it reached the dead end of a conscious mind encrusted with self-deception. Sustained conscious insight is not an option for the human animal. Because it hinders rather than contributes to our reproductive success, natural selection has "disfavored conscious knowledge of motivation as a social strategy."[25] For our species, all roads lead to self-deception . . . and thus to unconscious communication.

7

Machiavelli on the Couch

Speech was given to man to disguise his thoughts.
—Charles-Maurice de Talleyrand

It would be very useful to know more precisely what types of input jog the Machiavellian mind into action. This would allow us to infer more about how the Machiavellian module works, and would make it possible for us to identify situations where it can be observed in action. Although I presented a few ideas about this in the last chapter, these are not nearly specific enough to be of much practical use. In this chapter, we are going to turn to psychoanalysis again for a clue about how to proceed, since we have not yet exhausted what it has to offer us.

Wading through the murky waters of the psychoanalytic literature is not for the fainthearted. These writings can be exasperating if one approaches them with even modest expectations of methodological clarity and scientific objectivity. Psychoanalytic writings contain a wealth of theoretical speculation and many fascinating anecdotes, but they are hideously lacking in hard data and in trustworthy research methods. Although the bulk of this literature is, to say the least, misleading, chancing upon a single good observation may provide just what we need to complete another section of our puzzle.

A useful place to start is by considering the question of what made it possible for Ferenczi to make his observations of unconscious communication, crude and sketchy though they were. Obviously, there must have been *something* about the setting that activated his patients' Machiavellian modules, or else there would not have been any unconscious communications for him to notice. If we can distill what that "something" was, this may shed further light on what sets the Machiavellian module in motion.

Undergoing psychoanalytic therapy is a strange experience. In the "classical" Freudian modality, patient and therapist meet at a set hour from one to five times a week, always in the same room. The room contains a couch, upon which the patient reclines, and a chair situated behind the couch where the therapist sits, beyond the patient's field of vision. The neutrally decorated room reveals little about the therapist's life, values, or interests. There are no third parties present, and nobody can look or listen in.

The verbal exchanges, if "exchanges" is the right word, are nothing like normal conversations. The therapist asks the patient to say *everything* that comes into his or her mind and promises that everything that transpires will be strictly confidential. The therapist spends most of the time listening silently, speaking only occasionally to offer "interpretations"—statements that purport to spell out the patient's concealed thoughts. Psychoanalytic therapists do not engage in social chat, ask questions, or give opinions. Refraining from saying anything that is personally revealing, they are also careful to avoid cajoling, coercing, or seducing their patients into any particular attitude, emotion, or course of action.

The psychoanalytic encounter is deeply paradoxical. It is simultaneously intimate and remote. Therapists invite their patients to disclose raw emotional realities that they may have never shared with another soul, putting the therapist in a position similar to a trusted friend, family member, or spouse. But this relationship is also highly asymmetrical. The patient is expected to be frank and open, lying supine and exposed to unyielding, microscopic scrutiny, while the therapist remains guarded, anonymous, and inscrutable, withholding all information about his or her feelings, interests, and attitudes. A vast amount of sensitive, personal information flows from the mouth of the patient to the ears of the therapist, but hardly any passes in the other direction. Consequently, a person undergoing psychoanalytic therapy is unable to gauge the therapist's feelings about them. Is she disgusted? Amused? Aroused? Is she brimming over with compassion or unspoken condemnation? Does the therapist's silence express sympathetic attunement or sheer boredom? The therapist says little and sits out of sight, so the normal verbal and nonverbal clues are unavailable.

How would you think the Machiavellian modules of people undergoing psychoanalytic treatment are likely to respond to this bizarre situation? One thing is certain, the therapist's behavior is not unconsciously interpreted as having anything to do with an activity called "psychotherapy." Psychotherapy is an artifact of modern-day culture, and must be meaningless to a domain-specific module for social cognition. What our conscious minds call "psychotherapy" can only be unconsciously experienced as a conversation in which extremely personal and potentially damaging information is shared about oneself, one's friends, and one's family. It is a secret, exclusive conversation. In short, it is a form of gossip . . . with a difference. As we have seen, gossip is normally reciprocal and cooperative, but this poker-faced listener not only gives nothing away, but also is hell-bent on uncovering everything that the speaker wants to keep hidden. Think about this biologically. Putting yourself in the position of a psychotherapy patient's Stone Age social mind will drive home that this arrangement cannot fail to be deeply threatening. The power to deceive is our main weapon in the struggle for social survival. Like it or not, without it, we are sheep in the company of wolves. Similarly, the power to read intentions from nonverbal expressions is our best safeguard against victimization by others. Without it, we are at their mercy. Imagine that someone is sitting behind you, observing you closely and intent on uncovering the thoughts that you want desperately to keep hidden, while tenaciously guarding their own privacy. Do you get a whiff of danger? The therapist, it seems, bears the earmarks of a social predator. Psychotherapists inform us that people are often "resistant" to treatment. An evolutionary perspective tells us why. So-called resistance is an adaptive response to a potentially hazardous situation.[1]

Everything that we have learned so far suggests that the unconscious mind-reading circuits of anyone placed in the role of patient in the standard psychoanalytic setting will likely sizzle with activity. Like a brilliant detective, the patient's Machiavellian module will make the most of every small clue to determine exactly what the therapist is up to. Ferenczi was so astonished by his patients' mind-reading skills that he toyed with the idea that they possessed paranormal powers.[2] Fortunately, we no longer have to grasp at supernatural straws to understand unconscious perception. The phenomenon *is* amazing, but it is completely natural. It is no more paranormal than the elaborate predatory dance of the portia spider or the precisely engineered life cycle of the rat tapeworm. Objectively speaking, we should not be astonished to discover that a module unequivocally devoted to mind reading outperforms a general-purpose device (our conscious social inferences) in speed and in accuracy. Immensely rapid, specialized unconscious modules are humming in the background of our minds twenty-four hours a day. We could not get along without them. We could not get manage if we had to consciously coordinate our bodily movements, choose words in a conversation, or laboriously parse streams of sound from people's mouths into choppy words and sentences. Fortunately, our brains come equipped with pre-installed cognitive software for these tasks, and the same holds true of our ability to understand the meaning of social behavior.

Rules of Engagement

Robert J. Langs was the first psychoanalyst to study unconscious perception and communication systematically in the consulting room.[3] Although many of Langs's theoretical, clinical, and empirical claims are questionable, others are of real scientific interest and deserve serious attention. Years ago, when I worked as a psychotherapist in the United Kingdom, I had the opportunity to test out many of these principles in practice, and found that some of them were consistently and strikingly confirmed. Of course, these were not formal, scientific tests: they were informal clinical observations and therefore extremely vulnerable to bias and distortion. As I have already noted, we do not have the luxury of a body of empirical research to draw on, so relatively informal data gathering, with of all the uncertainty that it entails, will have to suffice. We must rely on anecdotes to help us get beyond the need to rely on anecdotes.

People tell their psychotherapists many stories, stories about childhood, friends and lovers, family, adventures and misadventures, triumphs and defeats. Langs proposes that these stories carry unconscious messages couched in an indirect, figurative idiom. They covertly express patients' unconscious interpretations of their therapists, and more often than not, portray the therapists as social predators. This unflattering portrait makes sense in light of my earlier discussion, and it is wrong to dismiss it glibly as a product of emotional disturbance or overactive imagination.

According to Langs, therapists' handling of the "ground rules" of treatment, the partially implicit and partially explicit norms governing the way that therapists manage the clinical setting, are the main triggers for encoded narratives. He suggests that psychotherapy patients unconsciously have an ideal concept of how the therapist should behave, against which they judge the therapist's actual behavior. When it falls short of, or contradicts, the ideal, patients encode their unconscious perceptions of this by telling negatively toned stories that spell out what has gone wrong. For example, a therapist who is late for an appointment may be privy to a patient's angry account of how a slapdash accountant delayed submitting the patient's tax return. After a brusque intervention, their patient may respond with memories of getting roughed up by the school bully. After announcing the need to cancel an upcoming appointment, the patient just happens to recall how his father was constantly away on business trips and took his family responsibilities too lightly.

Langs does not make the trivial claim that people have unconscious ideas about how others should behave toward them. He asserts that, irrespective of our conscious preferences, we all unconsciously measure others using the same unconscious yardstick. Can this possibly be right? When we look at human behavior, we find that some preferences are learned, but others are deeply ingrained in our human nature. Some people prefer AC/DC to the Jazz Messengers, Roquefort cheese to Brie, or London to New York, but you will be hard-pressed to find anyone who would rather drink urine than lemonade. It is obvious that there are species-wide standards of desirable conduct to which we expect our fellow human beings to conform. Nobody hankers after being betrayed, exploited, or cheated. These standards of conduct amount to "universal species-specific solutions for dealing with those features of the social . . . environments that were constant for our ancestors."[4] If, as Langs suggests, these unconscious standards embody the innate cognitive structure of a module for social cognition, the cumulative outcome of millions of years of social evolution, they *should* be uniform and relatively invariable.

Uniform standards for assessing human behavior should produce consistent social responses. Langs stresses that patients' narrative responses to therapists' behaviors are remarkably consistent—so consistent, in fact, that it is possible to *predict* the patients' narrative themes. I am aware that this sounds outrageous: later in the chapter, you will have an opportunity to try it yourself. Coded narratives are consistent because they are the output of a part of the mind that operates in much the same way in each of us. The "deep unconscious system," as he calls it, was designed by natural selection to quickly and accurately size up other people and to code its potent observations in stories. Although I disagree with many of his specific claims, at bottom Langs's "deep unconscious system" seems identical to what I call the Machiavellian module.

Betrayal Times Four

Having briefly sketched the theory, we can temporarily set it aside to delve into the realms of practice. I will present four brief vignettes culled from actual psychotherapy sessions, each of which involves a patient's unconscious response to a violation of confidentiality. Apart from my presenting them with a slightly dramatic flair, the only changes I have made were to disguise the identity of the people involved. There is nothing extraordinary about these examples. Similar tales unfold every day in consulting rooms all over the world, although their significance is hardly ever recognized.

Sam, a child psychotherapist, approaches the therapy room with his young patient in tow. Ali is a bright, active seven-year-old boy whose parents escaped to Britain from Sierra Leone. A well-intentioned teacher decided that Ali's boyish exuberance was a sign of psychological disorder and referred him for therapy. Ali shoves open the door and wildly rushes into a room full of toys, strewn wildly around the floor from the previous session. The room is just off one of the main stairways in a grim, sprawling Victorian school building. When classes change and torrents of noisy children rush up and down the stairs, some bang on the door or try to open it to discover what is going on inside. Children using the therapy room often react with alarm to these intrusions into their privacy.

As the therapist closes the door behind them, Ali scrambles over to the sandbox. After playing rather listlessly with the sand for ten minutes or so, he decides to enact a drama. He does this every week. Each time the plot is different, but the characters are always the same. In this week's episode, John and Harry visit a public swimming pool. After a long swim, they return to the locker room to change. The relaxed, pleasant mood of the story now yields to the staccato rhythms of anxiety. After undressing, John and Harry discover that their clothes are gone, apparently stolen. Standing naked in the locker room, they also realize that a television camera mounted on the wall is broadcasting images of the two of them into living rooms all over England. In a crescendo of panic, they try to hide, but there is no place to hide. Rushing to the door, they discover that they are locked in. Sam is puzzled about what this could possibly mean.

Time is up and Sam escorts Ali back to class. Returning to the room to tidy up, he tries to understand the preceding session. Why did Ali enact the narrative about being exposed, naked to the world? Glancing toward the door, where the imaginary television was mounted, he notices something that makes it all jell. The school caretaker had removed one of the two locks from the door, which left a gaping circular hole about an inch and a half in diameter. Any child coming down the stairs could stop on the landing and peer through the hole to observe surreptitiously everything going on in the room. Ali's imaginary adventure had a message: he invented a narrative of exposure because he was *in reality* exposed to the eyes of others.

A few miles away, Zoe sits waiting for her thirteen-year-old patient, Jane, to arrive. She waits with a bad conscience. Three days previously Jane's mother had telephoned to inquire about the progress of her daughter's treatment, and although Zoe had promised Jane strict confidentiality she yielded to pressure and discussed the child, albeit briefly. Did Jane know that Zoe betrayed her confidence? Would she talk about it in the session? Lost in thought, Zoe is startled when the buzzer sounds, and gets up to let Jane in. Jane sits down and begins the session by recounting an upsetting incident that had recently occurred at school:

> *Jessica . . . has now got her mother to ring up the school so that the head teacher could speak to Julia and me. She wants us to be nice to Jessica. I don't think she should have done this behind our backs. No wonder we can't be best friends. I thought I could trust her . . . When I was little, I used to go up to my room and lie on my bed talking to my teddy bears. I liked talking to them because I knew I could trust them. I could talk to them and they would not tell anyone what I had said because they could not speak. I could cry with them.*

Jane unconsciously addressed the conversation between her therapist and her mother by telling a story about a conspiratorial telephone conversation. Unlike therapists, it seems, teddy bears do not betray one's confidence.

A young man sits in the waiting room at a university counseling center in the heart of London. The building is busy and noisy and the room is poorly soundproofed. When Jeremy first came to the university medical center, the receptionist asked for his name, contact details, and a brief description of the problem for which he was seeking help. There were two guys sitting in the waiting room at the time, and although they didn't seem very interested, they could overhear everything he said. He then had to fill in a form, which the receptionist placed on top of the big stack of papers on her desk. Jeremy did not like the fact that two strangers had overheard his private business, but he shrugged off his feelings, reproaching himself for being oversensitive. He had a passing anxious thought about his personal information lying on the desk for anyone to see. "Your information will be kept entirely confidential," the receptionist chirped. "Only members of the medical center team will be able to read it."

Jeremy did not speak about his depression in his first session. After a few pleasantries, he addressed the therapist softly, with downcast eyes.

> *You know, in a place like this it is difficult to know who you can talk to . . . You don't know who you can trust. I was having a chat in the refectory with someone from my class the other day, when I noticed that two people I don't even know who were sitting at the next table were actually listening in to our conversation. My first reaction was to tell them something like "Why the hell don't you mind your own business," but then I thought, "What the hell."*

Was it coincidental that Jeremy spoke about eavesdroppers? Although he had consciously dismissed his feelings of discomfort about the lack of privacy at the clinic, his opening narrative told quite a different story.

Laura, a trainee psychotherapist, turns on her tape recorder and places it out of sight before ushering her patient into the room. Laura resisted doing this until now, because she had doubts about the ethics of surreptitious tape recording. Nevertheless, her clinical supervisor insisted that this is the best way to learn, and told her that there is no point burdening patients by telling them about it. "Anyway," Laura thought to herself, "this is a free clinic. The people here can't afford to see qualified psychoanalysts and if it weren't for students doing their internships, they'd just be given drugs." Laura opens the door to the waiting room and invites Mary to enter. Laura has been quite worried about Mary. Although she is in all other respects a practical, down-to-earth woman, she has recently developed some strange ideas. "I know it sounds crazy," she says, "but I can't help feeling that someone is watching me . . ."

> It's especially bad when I'm on my own. I walk from room to room expecting to see someone. It's creepy—I can't get over the feeling that there is some kind of invisible presence. I can't relax in my own home, there's always a sense that "It" is around. I don't like to have private conversations on the telephone, because I'm scared that "It" is listening in.

Could Mary have known about the concealed tape recorder? We do not have reason to assume that she did, as there is no reference to a tape recorder in her narrative. However, it is quite possible that Laura's mixed feelings about the taping leaked through her studiously professional persona, and that she involuntarily showed her hand, behaving in a way that indicated that they were not alone. Mary may have unconsciously read her tells. Remember, if the Machiavellian module exists, it must be far more adept at making this kind of inference than the conscious mind.

In each of these examples, there is a clear relationship between the breech of confidentiality and the resulting narrative themes. Each narrative is wonderfully distinctive in the way that it represents specific features of the triggering incident. In the case of Ali, the image of the video camera served as an apt analogy for the hole in the door through which people could look. These circumstances were very different from those found in the example involving Zoe and Jane, where a story about a video camera would have been entirely out of place. Instead, a sneaky telephone conversation was the narrative proxy for the therapist's chat with the patient's mother. In the third example, Jeremy was primarily concerned about being overheard and about his personal details being read by strangers and, sure enough, he told his therapist a story about eavesdropping. Finally, Laura's act of secretly tape recording her patient elicited the image of the intrusion of an invisible presence into private territory.

Predicting Narratives: An Exercise

Some years ago, I had an unusual opportunity to test some of these principles. I was one of five people invited by Michael Jacobs, a well-known British therapist, to write a commentary on a transcript of a tape-recorded psychotherapy session. Michael thought that it would be instructive to compare how commentators working from diverse theoretical perspectives made sense of a single session. He planned to publish the session transcript, the five commentaries, and some additional material as a book. "Ruth," his pseudonymous patient, gave her informed consent.

Before presenting a few excerpts from the session, I would like to invite you to try to predict Ruth's narratives. Many—perhaps most—readers will probably balk at the idea. How could they possibly predict stories told by someone about whom they know next to nothing? The fact is, if the theory is correct, you do not *need* to know anything other than the situation confronting her to which her Machiavellian module must respond. To predict the themes expressed by her narratives you need to think carefully about the main properties of Ruth's situation: she is about to be exposed to multiple third parties at her therapist's behest. Think of a few stories that could metaphorically express the raw Machiavellian implications of this particular situation. It may help to bear in mind that Langs claims that breaches of confidentiality are typically represented by stories involving betrayal, exposure, exploitation, assault, and sexual abuse.[5]

Here is what actually happened. The session began with relatively little narrative discourse, but somewhere near the middle, Ruth launched into a sustained episode of storytelling. Although there is another narrative episode later in the session, this sequence is by far the most vivid and emotionally potent. It begins with Ruth describing a recent experience of surgery.

OK, it was an intrusive procedure, and I knew what would happen then. But, even though I was expecting that—I had no idea what position I was on the operating table, or whatever—but to a certain extent, I don't know, but I would expect it to be intrusive. I am pretty certain that I must have been naked as well, because even though I had a gown on when I went in, when I came back out again it was done up differently. And during the operation I had been attached to a heart monitor. Consequently all the sticky things were over my chest and breasts, and that really hurt me. I mean not physically. I did not know what was going to happen, so I suppose to a funny extent I expected to be covered from here to here [neck to waist] and the fact that I wasn't just made me feel very, very cold . . . cold in the way of being . . . exposed and wanting to withdraw. That sort of cold.

Ruth appears to be representing the exposing qualities of the impending publication of her session and its circulation to the five commentators by an image of lying naked on an operating table. Notice also the reference to the heart monitor, which may be a disguised representation of the tape recorder. This is strong medicine, but there is more to come.

> *Yes, I had a lot of feeling when I came out and realized what had happened. It was going straight back to the accident, and obviously being brought in after the car crash and being stripped. I think that it was that that really . . . made me so vulnerable . . . I have no idea how many there are in an operating theatre, but it was almost as if it wasn't Jim who is the surgeon, it was everybody else who was there.*

The unknown strangers in the operating room are probably analogically coded narrative images of the strangers who are about to become involved in her private affairs. Ruth is naked and cut open for everyone to see.

You will need some background information to understand Ruth's next remarks. Earlier in her treatment, Ruth remembered a horrific experience of being raped by a gang of her brother's friends when she was only eight years old. Although many psychologists question the reliability of such "recovered memories," the question of their truth or falsity has no direct bearing on their *analogical* role. For our purpose it is the *story* that is most important.

> *It was that side I think that led me straight back to the abuse, and that it wasn't one, it was more than one. It just seems to be all . . . all muddling up. I was out of control again. I didn't know what was happening. That's what was making me sort of cold. . . .* [6]

The involvement of five commentators is now depicted as a gang rape, with all its implications of callousness, abuse, and exploitation. The narrative image of her brother's friends serves as a proxy for her therapist's colleagues. The string of images that Ruth presents are intensely disturbing: she is naked and exposed to many sets of eyes, her body suffers an intrusive procedure, and a gang penetrates and abuses her, pleasing themselves while she suffers. Of course, in reality, having five people write commentaries on your therapy session is nowhere near as terrible as being raped. This takes us back to the point that I raised in chapter 7 about the hyperbolic character of unconscious messages.

Dr. Jaws

The examples that I have presented so far all illustrate unconscious analogical communication, but fall short of supporting the claim that the Machiavellian module is a specialist at inferring undisclosed motives. Therefore, with this in mind, I will give a final clinical vignette before closing this chapter. Again, this is an account of a real psychotherapy session. The characters are Eileen, a female psychotherapist in private practice, and Nick, her male client.

After six months or so of treatment, Eileen began one of Nick's sessions by recommending that he meet with her more frequently. She suggested that they increase their meetings from once to twice a week, explaining that because Nick was wrestling with some particularly difficult, painful issues, the extra support would help him work through them more effectively. Nick had no reason to contest this suggestion. After all, Eileen was a seasoned expert and he believed that she had his best interests at heart. He spent a moment silently working out whether he could afford the additional expense, and then agreed to the proposal.

Having settled the matter, Nick began to tell his therapist about the weekend. He had not gone out Saturday night, but had stayed at home and watched the movie *Jaws* on TV. Had she ever seen it? Nick went on to tell her that the film is about a giant shark that terrorizes an island. He described the plot of the film in detail, with emphasis on the terrifying predatory greed of the shark. Eventually, he seemed to exhaust this subject and switched to something different. He spoke about driving through a seedy part of town on his way home from his psychotherapy sessions and mentioned that he often saw hookers soliciting men. He sighed and remarked how sad it is that these desperate women sell themselves on the street. Nick mentioned that there are plenty of beggars on the street there, and drug pushers too. At this point Eileen, who had been silent for most of the time, announced that time was up, and that she would see him on Tuesday at four, as they had arranged.

Eileen's intervention at the beginning of the session triggered Nick's narratives. When she proposed to Nick that they double the frequency of his sessions, and explained that this was in his best interests, Nick's conscious reaction was typically naïve. Yet again, we see the extent to which the conscious mind, clouded by self-deception, has a very limited grasp of the nuances and complexities of the social world. Compare this with the voice of his Machiavellian module, which tells a very different and much more sinister story.

Eileen made an ostensibly altruistic offer to Nick, offering to increase the frequency of sessions *for his own good*. Altruistic offers from people who are not blood relations, or with whom one does not share extensively overlapping interests, are biologically unrealistic and therefore suspect. This alone would be sufficient to activate Nick's Machiavellian module. Long ago the British psychoanalyst Margaret Little admitted in a very insightful and, at the time, controversial paper that psychotherapists sometimes try to *prevent* their clients from getting better. "Unconsciously," she wrote, "we may exploit a patient's illness for our own purposes. . . ."[7] Little believed that the main motive for such behavior was to provide therapists with a raison d'être; but, as we shall see, there are other, more prosaic, considerations to take into account. The Machiavellian module does not take altruistic gestures such as Eileen's at face value. A plausible answer to the question, "What's in it for her?" is evident from the themes that unfolded in his discourse. The first image characterized Eileen as the very antithesis of a caring altruist: she was seen as a greedy predator, a "shark." Why should this be? What Eileen did *not* mention in her intervention to Nick was that the increase in sessions would also be an increase in her income. The story of the shark indicates that Nick's Machiavellian module got wind of a concealed self-serving motive. This line is further developed in the tale of the prostitutes, who desperately sell themselves. Nick's Machiavellian module experienced Eileen as propositioning him, as whoring for business. What about the dope pushers? Well, this one also has financial connotations, but it may also encode an additional insight. It may be that Nick unconsciously concluded that Eileen wanted to make him dependent on her. This is not as far-fetched as it might seem. Dependence is one of the hazards of intensive psychotherapy. Many therapy addicts remain in treatment for years with no discernible improvement in their condition because they need their weekly "fix."

A Recipe for Mind Reading

We have to treat anecdotes from the world of psychotherapy cautiously. Nevertheless, the material canvassed in this chapter may add significant detail to our emerging picture of the Machiavellian module and its impact on human behavior. The most important lesson to be learned from it concerns situations that are likely to spur the Machiavellian module into activity. If my analysis is correct, people are most likely to tell unconsciously significant stories when:

   a. there is ambiguity or uncertainty about another person's stance regarding

b. covert conflicting interests, especially when expressed through

c. transgression or modification of an implicit or explicit social rule, particularly if

d. it is likely to be disadvantageous for the speaker to address the conflict directly.

It would be silly to claim that this recipe is complete, exhaustive, or even necessary for unconscious communication to take place. However, all of the illustrations used in this book, and many more, satisfy the recipe to a greater or lesser degree. Remembering the recipe can help one spot encoded narratives in daily life by alerting the person to situations where Machiavellian modules—including one's own—are likely to start churning out encoded narratives. It also opens the way for empirical research. In principle, psychologists could engineer situations satisfying the four conditions to stimulate unconscious communication. I use the circumspect phrase "in principle" because psychologists have not yet accepted that there might be a phenomenon here worth researching. My suggestions, however reasoned, are so far outside the mainstream that they are regarded as only slightly more plausible than reports of little green men in flying saucers.

Why Ground Rules?

The idea that the Machiavellian module is especially sensitive to rules and their violation can advance our analysis in interesting ways. All human societies have rules to structure the conduct of their members, and even wild chimpanzee troops may have rudimentary forms of them.[8]

Unless there is a compelling crisis, such as an attack by a common enemy, that creates a transient unity of purpose, the genetic differences inevitably create tensions leading to clashes over resource acquisition, dominance, sex, and so on. The absence of social rules would therefore give carte blanche for untrammeled coercion, manipulation, and interpersonal violence, to the detriment of everyone. It is only by limiting individual interests that human social systems can be sustained.[9]

Human beings have, and must have, an extraordinarily ambivalent attitude toward rules for social conduct. We are attracted to them because they protect us from abuse by others, but we also find them irksome because they make it difficult for us to promote our own interests by exploiting other human beings. In other words, we are prepared to accept the restrictions imposed by social regulations only if there is a hefty personal payoff for doing so. We piously proclaim that members of our reference group—our class, race, nation, and so on—are "good" people, to bamboozle them into keeping their self-interested inclinations on a tight rein, so that we have ample room to exercise our own.[10] To pull this off, most of us delude ourselves into believing that we are deeply committed to the ethic of justice and fairness for all. As Mose Allinson so gracefully put it, "Everybody's talkin' about justice, just as long as I get mine first." Rules and regulations drive self-serving motives underground. When our desires conflict with the established order, we pursue them secretly, deceptively and, as we have seen, often self-deceptively.

Groups are notoriously conservative, perhaps because attempts to bend the rules or create new ones are considered suspect by the majority. A person who advocates changing the rules is often felt to be promoting his or her interests at the expense of others. As Richard Alexander notes, "It is a common social strategy to assemble as a coalition those who agree, or who can be persuaded to behave as though they agree, and then promote the apparent agreement of the subgroup as gospel."[11] Natural selection tuned our Machiavellian antennae to detect rule transgressions because they signal social conflict. With a bit of forethought, we could have predicted the rule-sensitivity of the Machiavellian module on theoretical grounds.

8

Conspiratorial Whispers and Covert Operations

Poetry must be made by all and not by one.
—Isidore Ducasse

Why do we tell encoded stories? If our unconscious social intelligence existed only to inform us about our interactions with others, there would be no point in broadcasting the results to others. In fact, it would be disadvantageous; in the game of social poker, the less other people are aware of how much you know about them, the better. Our calling these narratives unconscious *communications* is already rich in implications: communication requires both a message producer and a message consumer, so the stories that we tell must be destined for other ears. But who is supposed to hear them, and why? My task in this chapter will be to tackle both of these crucial issues.

Whispers and Advertisements

Sometimes one can learn quite a bit about a message just by the way it is delivered. John Krebs and Richard Dawkins, whose work on manipulation and mind reading was introduced in chapter 2, point out that a loud, flashy delivery has a different set of implications than one conveyed *sotto voce,* inviting us to "contrast the Bible-thumping oratory of a revivalist preacher with the subtle signals, undetected by the rest of the company, of a couple at a dinner party indicating to one another that it is time to go home." Nonhuman species have their own version of hellfire and brimstone preaching called "ritualized signals" or "displays." Just like the ads on TV for the latest car or toothpaste, animals use "redundancy, rhythmic repetition, bright packaging, and supernormal stimuli" to manipulate others. Co-operating animals with confluent interests behave more like the couple at the dinner party. They signal discreetly, engaging in hushed conspiratorial whispers instead of the razzmatazz of ritualistic song and dance.[1]

Exploitative signals are brash and repetitive and cooperative signals are subtle and economical for good reason. Ritual displays owe their conspicuousness to the circumstances of their evolution: they are the outcome of evolutionary arms races. If a clever animal of type A is out to take advantage of a gullible animal of type B, those Bs that are able to resist A's sales pitch will be more successful than their more naïve peers. As the B population shifts toward greater resistance to A's manipulations, those members of A that are pushier than their comrades will melt the resistance of even the wised-up Bs, and so on. With each round of the tournament, A's "sales pitch" gets amplified and is eventually hyped into a loud, repetitive advertisement. In the case of cooperating organisms, natural selection shifts the process in the opposite direction; the volume of the signal is turned down to make it as unobtrusive as possible. In nature, every benefit comes at a cost. Brazen, conspicuous signals are more expensive than soft, subtle ones. Grand displays devour time and energy, attract the unwelcome attention of predators, and get in the way of performing other useful activities. An animal engaged in cooperative signaling, where the receiver welcomes the message instead of fighting against it, avoids these costs by reducing the conspicuousness of signals to a bare minimum. In this scenario, natural selection enhances the receiver's *sensitivity* to signals instead of boosting the volume of the signal itself. This is why skillful seducers in all areas of life cloak their persuasive efforts in the honeyed tones of cooperation.

Now, let us apply this framework to what we know about unconscious communication. Are encoded narratives loud advertisements or conspiratorial whispers? The four telltale characteristics of ritualistic signaling are redundancy, conspicuousness, stereotypy, and alerting components. If unconscious communication is an effort to persuade, it should possess most of these qualities, and if it is an effort to cooperate, it should lack them.

1. *Redundancy:* There is plenty of overkill in ritualized signaling (think of the highly repetitive content of TV infomercials). Unconscious communication is very economical, condensing a great deal of information into relatively few narrative images.

2. *Conspicuousness:* Ritualized signaling is full of intensity and sharp contrast (think of an orator at a podium, making a dramatic speech). Unconscious communication is easy to miss, even when you are looking for it.

3. *Stereotypy:* Ritualized signaling makes use of standardized components (think of Adolf Hitler's well-rehearsed repertoire of gestures). Unconscious communication is highly flexible and creative, selecting pertinent narrative images as the occasion demands.

4. *Alerting components:* These are conspicuous warnings that something important will follow (the tapping on

a wineglass before an after-dinner speech). We tend to slide in and out of unconscious discourse without warning, and it is difficult to determine where it begins and ends.

Coded narratives are obviously more like conspiratorial whispers than high-profile advertisements, which implies that they are cooperative signals. Given that I have spent the last three chapters arguing that the Machiavellian module evolved to manage deceit and counterdeceit, this result may strike you as more than a little disconcerting.

Unconscious social intelligence is a faculty for detecting cheating and social predation, but unconscious communication clearly has some additional purpose. Perhaps the relationship between them is like the relationship between predator detection and alarm calls in other species. When a vervet monkey spots a leopard in the vicinity, a distinctive pattern of visual input activates the monkey's leopard-detecting module, which triggers a distinctive signal to the rest of the group to take cover. The leopard-detecting module was a product of the evolutionary arms race between vervet monkeys and leopards. The alarm call, on the other hand, is a conspiratorial whisper to protect friends and family (the vervet alarm call is a "whisper" although it is loud—when whispering to a whole group, one has to be loud). This way of looking at things suggests that unconscious perception evolved in the context of an evolutionary arms race between conspecifics. This is completely consistent with the picture that I painted in earlier chapters. The novel element is the idea that the adaptive function of encoded communication is to warn one's allies of the presence of a social predator.

Before moving deeper into this analysis, we need to make sure that we are clear about some basic principles. Imagine that

Zara, an intergalactic archeologist from the planet Zongo, is excavating a site on an obscure, previously inhabited planet called "Earth" and uncovers a strange metal box with a transparent front and a cord extending from the rear. We earthlings would immediately identify the artifact as a microwave oven, but the Zongolians do not have anything like microwave ovens on their planet. In fact, they do not consume food as we know it—their nutrients are gaseous and absorbed through lung-like structures, so they have no concept of cooking and, by the way, do not use wires to conduct electricity. Unfortunately, no records survived the nuclear holocaust that wiped out the human species, so there is no documentary evidence for her to draw on. How does Zara make sense of the object?

The long and short of it is that unless she has some conception of its function, she could not get very far. She could analyze its form and composition, take its measurements, weigh it, and examine how its various parts fit together, but this would tell her nothing about what kind of object it is (A front-yard ornament? A terrarium? A space-age flowerpot?). When I use the term "function" in this context, I am not talking descriptively about what a thing *does*, rather, I am saying something about what it's *supposed to do*. Philosophers call this the "proper function" of a thing, as distinct from its plain old function. The computer monitor that I am looking at right now could be used as ballast to steady a schooner: that would be its function, but not its proper function.

Without knowing the proper function of the microwave oven, the alien archeologist could not even begin to consider how it works, because she would have no idea what it would mean for it to work. She would be unable to determine whether it worked well, poorly, or not at all. Finally, it would be impossible for her to infer its function, unless she understood the environment to which it was adapted (environments with electrical outlets, in which people cook their food, and so on). Biological systems, including mental modules and forms of behavior, are artifacts designed by natural selection. When we study them, we are in much the same position as Zara. To understand a biological system, we have to know what it is supposed to do, and we cannot know this unless we have a grasp of what kind of environment it was designed to plug in to.

We never diagnose the proper function of a biological system based on what it actually *does*. There are plenty of biological systems that only rarely fulfill the purpose for which they were selected. Think of the sperm cell: its raison d'être is to fertilize an ovum. And yet, the percentage of sperm cells that actually manage to pull this off is vanishingly small. Finally, a biological system needs to be in the right sort of environment—the environment to which it is adapted—to do what it is supposed to do in the way that it is supposed to do it. Sperm cells in normal environments rarely realize their biological destiny function; those in spermicidal environments (say, inside a condom) never do.[2]

If we are going to work out the proper function of unconscious communication we need to know something about the environment in which it evolved and to which it is adapted. The relevant archeological evidence is scanty, and tells us little more than that our ancestors were nomadic hunter-gatherers who periodically migrated in response to climate changes, who sometimes cared for their injured, and sometimes buried their dead. This bare description does not give us anything approximating the detailed understanding that we require. It can be enlarged somewhat using an approach called "cladistic analysis," which involves drawing conclusions from relevant similarities between closely related species. Around twenty million years ago the gorillas diverged from the chimpanzee/hominid line. Around fourteen million years later, the tree branched again as the chimpanzee-bonobo lineage and our own forked in different directions.[3] If we compare our social behavior with that of our nearest relatives, the chimps and bonobos, we find some interesting similarities. All three species compete for status, and tend to dominate and submit to dominance to form coalitions in the pursuit of power. All three also experience social conflict, which they appear to find distressing, and have methods for deliberate conflict resolution.[4] The fact that the social behavior of chimps, bonobos, and modern humans share these common denominators suggests that these were characteristics of our common ancestor. Extrapolating forward, it is a safe bet that our hominid ancestors also possessed them.

There is a way to push the analysis even further: we can use the present to draw conclusions about the past. If our remote ancestors, the earliest anatomically modern humans, lived as nomadic foragers, it is reasonable to think that their way of life had a lot in common with nomadic hunter-gatherers still around today. Although scattered all over the globe, and in many ways very different from one another, these groups have some striking similarities in their social organization. Nomadic hunter-gatherers tend to be moralistic and have very similar concepts of deviant behavior, including rape, theft, murder, deception, and the failure to cooperate. All use gossip to circulate information about transgressors, and have methods for identifying and punishing deviants, such as shunning, ostracism, ridicule, desertion, and in extreme cases, execution. Relations between adult males, and sometimes females, are highly egalitarian. Nomadic hunter-gatherers disapprove of and have found ways to suppress efforts by individuals to dominate and control others and therefore tolerate only minimal levels of leadership. They regard dominance as antisocial, with the whole group opposing anyone who tries to be a "big shot." Treating dominance as a

kind of deviance sharply distinguishes the political life of mo-
bile foragers from the social order of the chimps and bonobos.[5]

It would be a grave mistake to draw the sentimental conclusion that these people are "noble savages," innocent of any
thirst for power. In fact, warfare and homicide are quite common among them.[6] Wandering hunter-gatherers have not tran-
scended the urge for power: they found an efficient way to
police it. This is *counter*domination: domination by the majority
of group members rather than by one individual or a small
clique. All of this changed when our ancestors traded in forag-
ing for a sedentary lifestyle. Societies became much larger as
well as much more organized and stratified, and the accumula-
tion of wealth and, in particular, the establishment of standing
armies supported by wealthy rulers, sealed rigid dominance hierarchies and established extremes of political inequality. Iron-
ically, as social organization advanced in complexity, we gave
freer rein to the despotic, chimpanzee-like side of our nature.[7]

The evolution of language was crucial to the transition from
chimpanzee despotism to human egalitarianism. It permitted
literal conspiratorial whispering, enabling group members to
"arrive privately and safely at a negative consensus about dangerous deviants."[8] The Stone Age gossip networks were vital for
affiliation, as described in chapter 6, but were also crucial for
enforcing the ground rules of the group. Of course, there is always the question of *which* group. In the final analysis, a group
is a collection of people with interests in common. A renegade
clique will experience the rest of the group as deviating from
their conception of what is right.

Looked at in this context, it may be that unconsciously
loaded storytelling was initially a way of covertly identifying
social deviants. These narratives were not directed toward the
offender, but toward third parties, other members of the group

with whom the speaker was allied. Encoded narratives were a cryptic form of gossip, allowing members of the group to arrive unconsciously at a consensus about individuals in the group who pose a threat. This fits nicely with the concept of unconscious communication as conspiratorial whispering, and with the principle that the Machiavellian module is especially sensitive to transgressions of rules. The fact that far-flung hunter-gatherer groups share a consensus about what constitutes deviance, which was probably also shared by our Stone Age ancestors, may throw light on the idea that the Machiavellian module assesses human behavior according to a universal set of standards.

This model elegantly fits the vignette presented in chapter 6 about three students who turned up for class one winter morning to discover that the other four class members had remained in bed. Michelle's narrative branded the absent students as criminals, and she seemed to be unconsciously feeling out the other students' attitudes to the situation. It is also consistent with the other examples that I have given, such as the episode on the London bus, and the two stories of reactions to my interracial marriage to a much younger woman.

While this neatly explains what was going on in these encounters, it does not fit the examples from psychotherapy sessions presented in chapter 7. In these, there were no third parties present in the room and the unconscious message was apparently directed toward the therapist, who was experienced as deviant or predatory. These counterexamples are not unduly troubling because, as I pointed out earlier in this chapter, a psychobiological faculty can only fulfill its proper function if it is in the kind of environment to which it is adapted. The psychotherapeutic setting has features that trigger the Machiavellian module, but this contrived, deeply unnatural relationship is significantly different from the Stone Age social environment in which the Machiavellian module must have adapted. Consequently, it does not function properly.

All of this is speculation, but there are several ways that it can be tested. For example, the idea that encoded narratives are conspiratorial whispers implies that they should occur far more frequently in conversational groups of three or more people than in pairs. It also implies that the volume of these communications should be the minimum necessary for everyone in the group to hear, and we should find that encoded narratives should rarely if ever be loudly spoken. Finally, unconscious communication should be common in present-day social situations that are relevantly similar to prehistoric ones, namely small groups with minimal structure. The theory tells us we will discover that group members are very concerned, at an unconscious level, with deviant members. We should also find that *regardless of their conscious attitudes* group members are extremely wary of individuals who are dominant, or who try to become dominant, and will unconsciously spell out their reservations in encoded narratives.

Listening to the Subliteral

Robert E. Haskell, a cognitive and social psychologist and a colleague of mine, has spent over thirty years methodically investigating unconscious communication in small group settings. His core observations came from experiences conducting group-dynamic workshops known as "Training groups," or "T-groups," for short. These groups help participants learn about group processes by immersing them in a group experience.

The structure is simple: ten or so people meet at regular intervals to observe and discuss their behavior in the group. The label "training group" conjures up images of a trainer providing didactic instruction, but in reality, T-group meetings are quite unstructured. Trainers are nondirective, largely silent, and confine themselves mostly to occasional observations about or interpretations of the social processes unfolding in the group. The atmosphere is permissive, with very few rules, although self-disclosure, frankness, and a continual focus on the here-and-now are expected of members.

It is obvious that the T-group format has several features in common with psychoanalysis, such as the permissiveness of the setting, the mind-reading orientation and nonreciprocal stance of the trainer, and the emphasis on self-disclosure by group members. It is also easy to see that the T-group format puts its members in a double bind. Each person is to be open and honest about his or her feelings and perceptions of the others, while at the same time being intensely scrutinized by everyone else in the room. As we have seen, potentially hostile mind reading—mind reading between individuals who do not have significantly overlapping interests—*discourages* openness and promotes deceit. But deceit is difficult to pull off when ten people are watching your every move. This combination of pressures is a hothouse for self-deception and complicated Machiavellian tactics.

T-groups are rife with conflicting interests because members have no explicit common purpose except for the vaguely defined goal of studying how groups operate. Of course, the notion of studying group dynamics is utterly meaningless to the participants' Machiavellian modules, and although group members may consciously tell themselves that they are pursuing a common educational goal, their unconscious perspective on the situation is bound to be very different. The Machiavellian mind views the T-group not as an educational "group relations laboratory," but as an arena seething with primate politics, strivings for dominance, social predation, deception and all of the bluffs, feints, and ruses of advanced social poker. The ambiguous, nonreciprocal mind-reading role of the trainer is enough to awaken the unconscious social intelligence of group members, but the experience of being embedded in an unstructured group in which everybody is watching everybody else packs an extra wallop. T-group environment fulfills all of the conditions described in chapter 7 that switch on the Machiavellian module, and seems made to order for the study of unconscious communication.

During the early 1970s, at the same time that Robert Langs was formulating his ideas on unconscious communication, Haskell began to notice a strange phenomenon in the T-groups that he was running. Sometimes group members seemed to be speaking in a metaphorical idiom. The images and themes in the conversations that bounced around the group from one speaker to the next seemed to be veiled references to the group itself. Although much of the conversation in T-groups seems on the surface to be relentlessly trivial and apparently random chitchat, Haskell began to realize that it is underpinned by a deeper, more meaningful order.

These *subliteral* (or sub-literal) communications, as he began to call them, have a quality of immediacy. Unlike ordinary conscious discourse, which ranges freely through space and time, subliteral communications center on the here-and-now, on the struggles in the group with the group.[9] This process rolls along outside the awareness; group members have no inkling of the poetic, multilayered, analogical dimension of their conversations. For instance, when group members were preoccupied with group meetings being videotaped, they talked about the FBI and the CIA. When it became clear that the trainer would not share his notes with them, the conversation would drift to the topic of investigative reporters who refuse to reveal their sources. On one occasion, a passerby mistakenly opened the classroom door and, realizing his mistake, quickly closed it and left. A group member remarked unexpectedly that she once had a mouse that would peek into her garage at the animals that she kept there.

Haskell, a self-confessed monomaniac, immersed himself in a detailed study of group talk. He scoured the literature for references to the phenomenon, recorded T-group sessions, closely analyzed the written transcripts, and soon began to notice the same processes at work in ordinary conversations. "At this point," he wrote, "I found myself suspecting my observations—and my sanity. . . . After all, I observed that group members were talking to me in code, as it were. This is the stuff that paranoid schizophrenia is made of."[10] This was no idle worry. Thirty years earlier Ferenczi's colleagues, including his one-time mentor Sigmund Freud, wrongly diagnosed him as psychotic. They apparently thought that Ferenczi's idea that his patients were talking about him in coded messages sounded like it was a symptom of severe mental illness. It is important to realize that, as a psychologist, Haskell was working in a professional climate that was deeply antagonistic to the very idea of encoded, unconscious messages. Many scientific psychologists, both then and now, are chary of getting anywhere near anything that smacks of "Freudianism." These people usually do not know very much about what Freud actually wrote, but this does not prevent them from trashing or cold-shouldering anything that they think *sounds* Freudian. To his consternation, Haskell found that he was often tarred with this brush.

Haskell's observations of subliteral messaging in small, unstructured groups provide a wonderful opportunity to explore the hypothesis that unconsciously encoded stories are conspiratorial whispers. Along the way, I will explain some of his ideas about how unconscious communication works that are rather different from anything than we have covered so far.[11]

The Poetics of Deviance

When I was a child, my friends and I would sometimes amuse ourselves by collectively inventing a story. One of us would begin and, after a couple of sentences, the next kid would take a turn, and so on round and round the circle. Each of us tried our hardest to create weird, novel twists, so that the tale was constantly poised on the edge of incoherence. Later in life, I learned that a similar game called "*le cadavre exquis*" (the exquisite corpse) was played by French surrealists during the 1920s. Each player would write a phrase on a sheet of paper, fold it over to conceal their contribution, and then pass it to the next person in the group. The game got its name from the eerie sentence created the first time it was played: "*Le cadavre exquis boira le vin nouveau*" (The exquisite corpse will drink the young wine). The surrealists believed that this game tapped the unconscious personality of the group.[12]

Listening to subliteral talk in a T-group is like sitting in on a repeated game of the exquisite corpse. The conversation jumps from topic to topic, often in a seemingly random and chaotic manner, as a variety of images are woven into the collective story. Look a little more carefully, though, and you will begin to see that there is a strange harmony at work. Many of the narrative images and themes cohere meaningfully, and in very surprising ways, around a single topic. We do not have to resort to mysterious notions of a "group mind" or "collective unconscious" to explain this phenomenon. The stories fit together insofar as interests of the members converge and fly apart when they conflict. Because the unconscious social mind of every one of us is cut from the same biological cloth, we are bound to respond to the adaptive challenges of group life in similar ways.

To appreciate the flavor of group talk, consider the following extracts from the first eight minutes of a T-group session.[13] A student named Paula began the session by telling the others how she and some other students began a discussion in class but continued it in the student center. She described how another student reproached them for carrying on the discussion in this setting.

Paula: You know that discussion here last week . . . in Adolescence? We went over to the Center, you know . . . we were discussing it . . . And she came up and she said, "I think you're kind of juvenile for discussing it out of class."

After several short exchanges, the theme continued as follows.

Mark: Yeah, well she was just talking to me afterwards in the Student Center. She thought it was pretty rude the way she was in the bathroom or something and she heard some girls talking about how uh somebody was getting down on Madeline because she was uh saying that she didn't care about people smoking marijuana and they thought that it was so terrible because she was an older lady. And that's what I heard from Marge.

Why is Mark speaking about someone eavesdropping on gossip? To begin to think about this question we need to know something about the setting. The group is actually an undergraduate class on group dynamics. Group members know each other outside the group, and many of them share friends and acquaintances. The students have all agreed not to share information discussed in the group with outsiders, but of course, all of them know that this is unrealistic. There is always a chance that what one says in the group will end up getting passed through the university gossip mill. Gossip is, after all, precious social currency.

There are some other threats to confidentiality too. The session is being tape-recorded and there is a one-way mirror on the wall. For all the students know, someone may be observing them from the adjoining room. The instructor requires each group member to keep a journal to record their observations, and to submit papers to him on the dynamics of the group. Might he share these with others?

Paula: Yeah, she was . . . I mean like I said that . . . I just said that I didn't think that this discussion should have taken place because I thought they were getting awful personal.

Sandra: They were. Especially to one girl . . . and . . . it wasn't the students that were getting personal. It was the instructor. The instructor was coming out and just asking very personal questions. She says, "Oh, I didn't mean to get personal or anything, but you know, da, da, da."

Barb: I'm not saying . . . you know. I don't know Susan and I would never say anything against her because I don't know her. Mmmm. Well, she's lived next door to me all my life.

Paula: But I know we've just dropped a long way since. But I think she had kind of gall to come up to me and say that, especially in front of a group.

The instructor who asks inappropriately personal questions, and thereby exposes a student to a group of her peers, is perhaps a proxy for the T-group trainer. Sandra, Barb, Paula, and Mark seem to be in unconscious agreement that the T-group setting is too public.

Mark: It sounds like the instructor went a little bit too far. I think she was trying to get . . . She said she used to be a personal counselor or something.

Paula: Yeah and I think . . .

Mark: Like maybe she's bringing a little bit too much of that into the class. I like her. She's a good teacher, but she might be trying to bring a little too much of that into the class.

Mark: Like this diary thing that we're writing. Do you do that in your class?

Paula: No.

Mark: Diary? Not at all. That's sort of prying, but she says that . . . you know . . . nobody else can read it. Of course I'm being recorded right now. But that's . . . We have to depend on our professional status of our teachers. We had that in what? Skills and Methods?

The story about the instructor requiring them to keep diaries may stand for the trainer tape-recording the group meetings, the journals they are required to keep, and/or papers analyzing the group process that they are required to submit. Mark then directly mentions the tape recorder, and consciously and probably self-deceptively covers his tracks with the oddly bungled remark, typical of conscious naiveté, about depending on the professional status of teachers. Then he quickly gets himself off the hook by asking a question.

At this point, the trainer intervenes, pointing out some general similarities between the group's narratives and the class itself, and specifically connecting talk about the diaries with the group-dynamics journals. Mark's response is fascinating, as is the reaction of the rest of the group.

Mark: Exactly what we have been reading. That's why I put that in, that really in all classes we have to depend on the professional status of our teachers, just like as if they'd be a doctor or a dentist. They're not going to go and get drunk and use those tapes as entertainment for a party or something like that. You just have to trust they don't do things like that, and that a teacher wouldn't. . . . There are ways of having him removed.

[LONG SILENCE AND WHISPERS]

Mark has evidently got too close to the bone. The group falls silent and then members begin to whisper—presumably so their voices will not be picked up by the microphone.

Mark: Do you think our society has pushed that type of philosophy on us, with the constant media talking about . . . of, just recently we had that Right to Privacy Act passed. Do you think all that's made us a little bit paranoid about people's listening in to us, people opening up our letters, getting a dossier on us if we happen to be present at a demonstration? I really think that it has made us paranoid a little bit.

Sheila: Well, I think probably we had this feeling up until recently that all this was kept in confidence, and now that we realize that it isn't has made us more aware of the fact that what we say and do is open to criticism . . . from any angle really or from any number of things. Do you think so?

Kate: I don't agree with you.

Sheila: I think it's been magnified or probably brought out more clearly . . . The fact that we really don't have any privacy.

Mark: Mmmmm. Just recently, within the past five years I guess, maybe it's just me, maybe it's been known by people from all time, but I don't know, maybe it was just me that was awakened to it recently, the fact that like those Chicago 7, those trials and stuff like that I heard that if you showed up, well, it's a fact, that you show up for a demonstration such as that and it's very, very possible that the FBI or some organization has a dossier on you, because, you know, labeled as a radical. And then possibly if you go for some really prominent job and, maybe a CIA job or something like that, and they go back and go to one of those organizations and say, "Oh, this guy all-American red-blooded, you know, Yankee Doodler?" Then they say, "Oh well, he's participated in this certain demonstration." I don't know, I believe that could probably keep you back from the job.

The tape recording and other avenues for information leakage has led Mark to protest against the trainer, and even imagine getting rid of him. The "Chicago 7" were protesters at the 1968 Democratic National Convention, who were arrested and brought to trial. The FBI or CIA agents who keep secret dossiers on protesters stand for the trainer, who is keeping notes on them. Mark's story about someone being denied a position because they turned up at a protest demonstration, and his reference to the "Chicago 7" suggests that he is worried that the trainer may penalize or ostracize him. The odd term "Yankee *Doodler*" may obliquely refer to the fact that the trainer is taking notes; this would cohere with the theme of secret dossiers.

The literature on group dynamics recognizes that issues about leadership or authority are near the top of the agenda for unstructured groups. The most dominant person in the T-group setting is obviously the trainer, but the issue also extends to ordinary group members who inevitably vie for control over the course of the group, either singly or in politically astute co-alitions. In the preceding example, the session began with the group unconsciously expressing its concerns about privacy and confidentiality, which soon morphed into a challenge to the trainer's dominance, spearheaded by Mark, the only other male in the group. If encoded stories are a kind of conspiratorial whispering intended to unite community members against deviants and despots, we should discover that rank and file group members unconsciously disparage the trainer and anyone else who seeks a dominant role.

Judging from Haskell's descriptions, this seems to be what happens. Thumbing through his vignettes, we find the trainer unconsciously represented as a spy, a reporter who will not reveal his sources, and a doctor who writes unintelligible prescriptions. He is also unflatteringly portrayed as inadequate, overrated, out of touch with reality, a brainwasher, patronizing, dogmatic, and insulting, to name just a few of the many uncomplimentary characterizations. This disparaging attitude toward leadership also includes ambitious peers. "Peer leadership," he writes, "tends to be rejected or resented."[14]

Up until this point, we have decoded the unconscious content of a piece of conversation by extracting its themes. For instance, the theme of Michelle's story described in chapter 6 was something like "someone selfishly goes off to enjoy themselves and cruelly leaves someone else to die." It was easy to see a connection between this and the situation in class: the absent students had indeed abandoned their peers instead of honoring their commitment to the others. I also called attention to the point that the abandoned child was described as being *three* years old, and hinted at a possible connection with three students in class that day. If this was more than just a coincidence, it opens the possibility that a story's theme does not exhaust its unconscious meaning. While not neglecting broad themes, Haskell attends closely to the specific words chosen by the speaker. Number references, puns, and directional metaphors such as above or below have an important place in his approach to unconscious communication.

Now, let us continue looking at the T-group session and concentrate on the members' responses to leadership issues. For reasons that will soon become clear, it is important to know that this particular meeting consisted of the trainer, who was in his late thirties, Mark, who was twenty-two years old and the only other male, Sheila, a woman in her late fifties, and ten young women, who were all less than twenty-one years old. The two dominant members of the peer group were Mark and Sheila, who, together with the trainer, made up a dominant triad. As the meeting wore on, attention focused on the dominant members. Haskell counts no less than *fourteen* subliteral references to them. Rather than laboriously reporting all of them, I will describe some of the highlights of the session to give you the flavor of Haskell's more microscopic style of analysis, and of how unconscious discourse pans out in the group setting.[15]

Early in the session, one of the young women mentioned that "about three weeks ago" a group of police were "all down behind the Pantry Pride." The first thing to notice is her use of the number three. Could this be a subliteral reference to the three dominant individuals? The phrase "all down behind" gets its sense from the metaphorical significance of "down" and "behind," which have connotations of inferiority. So, "all down behind" may represent the subordinate status of the ten young women. What about the supermarket "Pantry Pride"? Sheila had previously remarked that she was *proud to be a homemaker*. "Pantry Pride" might be understood to be a disparaging moniker for the dominant member of the female subgroup, who were "down behind" her. Why would this young woman represent

the subdominant majority as police? This would be rather mysterious if we did not have the Stone Age social model to draw on.
The narrative image of a police posse may well refer to the
contra-dominant, policing function of group members.

Soon afterward, one of the young women told a story about
a bar in which most of the customers were "under twenty-
one." I have already mentioned that all ten of the subdominant
women in the group were actually under twenty-one years old.
However, there may be more here than meets the eye. According to Haskell's theory of unconscious communication, numbers behave differently in subliteral discourse than they do in
normal, conscious talk. In conscious talk, the phrase "under
twenty-one" refers to a *property* of the customers in the bar,
namely their age, but in subliteral discourse, numbers normally
stand for subgroups, cliques, and coalitions. Haskell suggests
that in this case, saying that the bar's patrons were "under
twenty-one" doesn't just mean "less than twenty-one years
old," it means "being subordinate to (under) the persons referred to as 'twenty-one'." To whom might this refer? The
number twenty-one breaks down to two (a two-person subgroup) and one (a single individual). So, being "under twenty-
one" means being subordinate to a coalition of three people,
two of whom form a subgroup.

I am well aware that this may sound like a pseudoscientific
number game. But before tossing it into the trash can of
charmingly bad ideas, recall that the discussion of Robin Dunbar's work in chapter 6 suggested that we should expect the
Machiavellian mind to be obsessed with subgroups, cliques,
and coalitions. Like any good political strategist, it has to monitor the various factions in its community, and track the size and
composition of each subgroup. There is nothing intrinsically
bizarre about a number-crunching unconscious mind. Although it seems strange that the Machiavellian mind would

represent subgroups in such a bizarre fashion, what counts is whether the relationship consistently and systematically holds. Do references to the numbers 12, 21, or 111 appear with greater frequency in groups with a triadic dominance structure than they do, say, in groups with two dominant individuals? Only well-designed empirical research can provide a definitive answer to this question.

The group spent quite a bit of time on the story about going to the bar, which culminated with an anecdote about ten people, three of whom were drunk, and the bartender who refused to serve them. Again, the number three seems to index the three dominant members of the group. But why would the subdominant speaker represent them as drunk? In ordinary vernacular, we speak about being "drunk with power," being "intoxicated" by it, and of power "going to your head," metaphors that may be based on the mood-altering effects of serotonin.[16] We heard another example in the introductory segment, when Mark talked about the trainer getting drunk and entertaining party guests with his tape recordings. The narrative theme of drunkenness was apparently used to represent an abuse of power.

The number ten may have been introduced because there were ten subdominant members present in the group that day. The remark that "they wouldn't serve any of them" ("them" being the three drunks) seems to turn on the word "serve." Subordinates *serve* their superiors, so perhaps the speaker was urging the rest of the young women to stop pandering to the three dominant individuals. As another group member went on to remark, "The bartender can refuse if he's served you, for instance, three drinks"! Yet another mention of the number three cropped up in a story about three high school seniors who got drunk on an airplane during their class trip, which was told by one of the young women. Notice that the three intoxicated passengers were *seniors,* which may denote the age difference between the dominant and subdominant group members, and note that they were *high* school seniors. Being "high" is a spatial metaphor for dominance ("high and mighty") and can also denote a state of intoxication ("getting high"). It should come as no surprise to hear that the intoxicated passengers were kicked off the airplane because of their bad behavior. This certainly sounds like a conspiratorial whisper, directed to the subdominant majority, about removing the three leaders. After a sequence of three more remarks also involving the number three, the next story involved a disparaging reference to three dilapidated buses. The speaker described the buses (leaders) as *old Grey*hounds, homing in on the status conferred by age, but also suggesting that they were well past their prime.

A Question of Method

The hypothesis that unconscious communication is a cryptic form of conspiratorial whispering is consistent with the data reported by Haskell. Of course, this is not nearly enough to put it on a sound footing, but it is a beginning; and, until other researchers become interested in putting unconscious communication to the test, it will have to do.

At this point I envision readers shaking their heads from side to side in amazement at my limitless credulity. Perhaps you are thinking that this stuff might be acceptable as literary criticism, or discourse analysis, or phenomenology, or psychoanalysis, but that it sure is not science. This is a familiar objection, which is sometimes followed with comments to the effect that the discourse of science should be kept separate from that of the humanities. As long as we give each of them equal dignity within its proper sphere, then everyone will be happy. But keeping everyone happy is not the aim of serious inquiry; its aim is to discover what is really going on in the segment of nature that one is studying. Finding out what's going on is the task of science, and not its humanistic alternatives. A less tactful version of the same objection might accuse me of wild, speculative, and ad hoc interpretations of the data. These interpretations *are* speculative, and some are much more so than others. This is inevitable, and I make no bones about it, but there is a large logical gap between this and the assertion that the whole approach is unprincipled. I am reminded of the maxim that "In the natural sciences a person is remembered for his best idea; in the social sciences he is remembered for his worst," which Richard Alexander, whom I am always in danger of quoting to excess, glosses as follows:

> *If a natural scientist were to write a long book in which a single error occurred, most of his peers (except, perhaps, his closest competitors) would probably feel that they could judge the error and the rest of the book independently. Conversely, if a social scientist were to write a long book otherwise of great value in which a sentence such as "Hitler was a great man" appeared, I am sure that everything else in the book would be discarded by most as suspect because of the single preposterous statement.[17]*

Unlike science, which has, at least somewhat, learned about the value of cooperation, social science is "red in tooth and claw." As Trivers perceptively notes, each school of thought in psychology is specialized in showing why the other schools of thought are mistaken.[18] Hostility to innovation is nicely illustrated by the example of a colleague whose research paper (which, I hasten to add, had nothing to do with unconscious cognition) was rejected by a major journal on the grounds, shamelessly stated by the editor, that her results were inconsistent with results produced by previous studies! Attitudes like this make it very difficult for a discipline to progress.

Haskell's interpretations are not "wild." They are based on the idea that there are consistent relationships between the verbal imagery used by group members and the psychological tensions operating within the group. In the examples mentioned above, Haskell did not simply latch on to the number three because it suited his purposes. His interpretations were guided by a rule stating that when a group is concerned with issues of dominance, the number corresponding to the number of dominant individuals will be selected into the conversation. In fact, he suggests that there is a whole network of such consistent relationships.[19] For example, when a number representing the dominant members of the group is introduced, it appears in the context of metaphors that denote dominance (for example, "on top," "up," "in front") and not subordination ("on the bottom," "down," "behind"). The status of the speaker also makes a difference. Members of dominant cliques imbed the number in positive contexts, while subdominant members give it a more negative spin. (A dominant group member might refer to the three wise men of the Bible, while a subdominant might mention the Three Stooges.) These claims may be hard to accept, and may be in error, but they are certainly not unprincipled.

It is quite likely that many of the hypotheses proposed in this book will eventually turn out to be inaccurate. However, if we rejected theories because they *might* be wrong, we would not have any theories left. The notion that our ordinary social talk resembles intricately constructed poetry, and reveals cognitive abilities of which we are completely unaware is certainly *strange*, but since when does strangeness count against a theory? Scientific theories are often weird, because they reach beneath the surface of the world in an effort to grasp its inner workings. Accepting or rejecting psychological theories because of their "resonance" with the way that you feel yourself to be plays right into the hands of self-deception. The evolutionary perspective on self-deception entails there are some psychological theories that will *never* feel right because they address aspects of human nature that nature designed our minds to reject. It is explanatory and predictive power that makes a theory scientifically meaningful, and we will not have a clear verdict on the ideas described and advanced in this book until the day that they come under serious empirical scrutiny. If this happens and they are not vindicated, I will finally be liberated from the grip of a compelling illusion that has enthralled me for the last twenty years. Let the chips fall where they may.

Descartes's Demon

*Why does man not see things? He is himself
standing in the way: he conceals things.*

—Friedrich Nietzsche

In the first of his famous *Meditations,* Descartes imagined trying to outwit an all-powerful demon that was determined to deceive him. Of course, Descartes did not really believe in the
demon. He used the idea as a gimmick to keep him awake and
on his toes. "I will take great care not to assent to what is false,"
he wrote, "nor can that deceiver—no matter how powerful or
cunning they may be—impose anything on me."

> *But this is a tiring project and a kind of laziness brings me back to
> what is more habitual in my life. I am like a prisoner who happens
> to enjoy an imaginary freedom in his dreams and who subsequently
> begins to suspect that he is asleep and, afraid of being wakened, con-
> spires silently with his agreeable illusions.[1]*

Descartes was after nothing less than a secure foundation for
knowledge. He thought that knowledge must be certain, and
that if we can call something into question we do not *know* it.
He therefore decided to pour the acid of doubt over everything
that he believed, to see what, if anything, remained undis-
solved. As it happened, not enough survived his assault to provide the secure foundation that he was seeking.

It was at this point that Descartes flinched. If only he could prove that he had been created by a benevolent God, then everything would be all right. In contrast to the elegant skeptical arguments in his first and second meditations, Descartes's "proofs" of God's existence are embarrassingly bad. Descartes lost his nerve. He stopped wanting to wake up, and hunkered down in a soothing dream of God and goodness and truth.

In the end, the demon got the last laugh. Descartes did not know the demon's name, but we do. It was René Descartes.

Over the last four hundred years, physical science has made magnificent strides. The contrast with social science would be comical if it were not sad. If the evolutionary account of self-deception is right, human nature stands in the way of understanding human nature. "In all the universe," writes Richard Alexander, "the only topic that we literally wish not to be too well understood is human behaviour."[2]

Self-deception has done us proud. Without it, we would probably never have developed our species' complex social forms and might still be running naked through the forest (which is not an altogether unappealing idea). Self-deception has been a wonderful gift, but it is now destroying us. Our taste for it resembles our craving for sugar and animal fat. These were good for our Stone Age ancestors, who had trouble finding enough nourishment to keep them alive and had to trek for hours each day to get it, but they are terrible for a sedentary, well-nourished population that need only drive down the road to get a fast-food fix. Self-deception was a splendid adaptation in a world populated by nomadic bands armed with sticks and stones. It is no longer such a good option in a world stocked with nuclear and biological weapons. The problem is, we are stuck with it. The switch is jammed and we cannot turn it off. It looks like natural selection may get the last laugh.

The most dangerous forms of self-deception are the collective ones. Patriotism, moral crusades, and religious fervor sweep across nations like plagues, slicing the world into good and evil, defender and aggressor, right and wrong. In the past, the mechanisms of group competition paid off. Self-righteous crusades vastly improved the quality of life for the victors and often for the defeated as well. Now, in the twenty-first century, it is likely that we will all be losers.

If we cannot transcend the demon, because we are the demon, we can at least try acknowledging its existence. Here's a novel suggestion: Let's start making a real effort to stop telling deliberate lies about human nature. The alternative approach no longer seems to be working all that well. Let's teach children real history instead of fairy tales, treat blind loyalty to a cause as shameful rather than virtuous, and pay as much attention to inoculating populations against virulent illusions as we do to inoculating them against contagious diseases. Let's get real.

I don't for a minute believe that we can be taught not to deceive ourselves, and even if we could (by whom?), it would probably result in widespread unhappiness. We are all frail creatures who need something to get us through the night. But surely, we can get rid of some of our surplus self-deception. Tolerating a measure of self-deception is one thing, but actively promoting it is quite another. At a minimum, perhaps we can help each other to acknowledge that we are all natural-born liars.

If the concept of the mind presented in this book is anywhere near to being accurate, human beings know far less about themselves, and far more about other people, than they are aware of knowing. There is an ancient joke that most psychologists have heard at least a dozen times about two behaviorists making love. After a thunderous climax, Fred lights a cigarette, turns to Mildred, and says, "It was great for you, but how was it for me?" The point of the joke was to poke gentle fun at the behaviorists' theoretical devaluation of human subjectivity and to show how absurd it would be to take this attitude out of the laboratory and into the bedroom. But the story contains a germ of wisdom of a different kind. In was in our ancestors' interests to deceive themselves about their own agendas and to develop an exquisite sensitivity to the mental states of others. To understand ourselves, to counterbalance our own self-serving biases, we need to look at ourselves in the mirror unconsciously held up to us by our fellow human beings. For those who value the advice of the Delphic oracle to "know thyself," the Machiavellian module is well worth listening to.

Unconscious Creativity

Many artists and writers have stressed the importance of un-bidden thoughts for their creative work. Samuel Taylor Coleridge described in the "prefatory note" to *Kubla Khan* that the poem was composed during a three-hour afternoon nap. Because he was disturbed by a "man from Porlock," Coleridge was able to record only a fragment of the three hundred or so lines that he composed in his sleep.[1] The French poet Paul Boux hung a sign on his bedroom door before retiring to bed on which was written: "Poet at work."[2] William Blake informs us that his long poem "Jerusalem" was written "from immediate Dictation twelve or sometimes twenty or thirty lines at a time without Premeditation and even against my Will."[3] A. E. Houseman records taking afternoon walks during which "there would flow into my mind, with sudden and unaccountable emotion, sometimes a line or two of verse, sometimes a whole stanza at once, accompanied, not preceded, by a vague notion of the poem which they were destined to form part of." "A common phrase among poets," wrote Amy Lowell, "is, 'It came to me.'" So hackneyed has this become that one learns to suppress the expression with care, but it really is the best description I know of the conscious arrival of a poem.[4] Mozart described the arrival of a piece of music in much the same way:

When I am, as it were, completely myself, entirely alone, and of good cheer—say traveling in a carriage, or walking after a good meal, or during the night when I cannot sleep; it is on such occasions that my ideas flow best and most abundantly. Whence and how they come I know not; nor can I force them.[5]

Richard Wagner composed the overture to *Das Rheingold* in his sleep. He wrote in his autobiography how he escaped the bustle of Venice to the sleepy village of Spezia. Feverish and unable to sleep, Wager took a long country walk the next morning:

Returning in the afternoon, I stretched myself, dead tired, on a hard couch, awaiting the long-desired hour of sleep. It did not come; but I fell into a kind of somnolent state, in which I suddenly felt as though I were sinking in swiftly flowing water. The rushing sound formed itself in my brain into a musical sound, the chord of E flat major, which continually re-echoed in broken forms; these broken chords seemed to be melodic passages of increasing motion, yet the pure triad of E flat major never changed, but seemed by its continuance to impart infinite significance to the element in which I was sinking. I awoke in sudden terror from my doze, feeling as though the waves were rushing high above my head. I at once recognized that the orchestral overture to the Rheingold, which must long have lain latent within me, though it had been unable to find definite form, had at last been revealed to me.[6]

Beethoven and Tartini had similar experiences. Paul McCartney first heard the haunting melody of "Yesterday" in a dream.[7]

Experiences such as those reported by Coleridge, Houseman, Mozart, and Wagner have also played a significant role in scientific invention and discovery. The Russian chemist Dmitri Mendeleev discovered the final version of the periodic table of the elements in a dream during a catnap. Mendeleev had been playing solitaire and laid his head down on the table for a short snooze. A dream about the tableau of the solitaire game supplied the form of the table in which he placed the elements upon awakening.[8] Another chemist, the Nobel laureate Melvin Calvin, was troubled by laboratory findings that conflicted with his conception of photosynthesis, until . . .

> *One day I was sitting in the car while my wife was on an errand.*
> *For some months I had had some basic information from the labora-*
> *tory that was incompatible with everything that, until then, I knew*
> *about the cycle. I was waiting, sitting at the wheel of the car, proba-*
> *bly parked in the red zone, when the recognition of the missing com-*
> *pound occurred. It occurred just like that—quite suddenly—and*
> *suddenly, also, in a matter of seconds, the cyclic character of the path*
> *of carbon became apparent to me. But the original recognition . . .*
> *occurred within a matter of thirty seconds. So, there is such a thing*
> *as inspiration, I suppose, but one has to be ready for it.[9]*

The physicist Leo Szilard was standing at a pedestrian crossing at a busy London intersection when "As the light changed to green it suddenly occurred to me that if we could find an element . . . which would emit two neutrons when it absorbed *one* neutron [this] could sustain a nuclear chain reaction," a "Eureka!" experience that led to the first atomic bomb. The eccentric Serbian inventor Nicola Tesla invented the principle of the rotating magnetic field in an unexpected explosion of insight while strolling through the park reciting Goethe's *Faust*. The experimental procedure that demonstrated the chemical transmission of nerve impulses came to Otto Loewi in a dream, as did the design of Parkinson's computer-operated anti-aircraft gun.[10]

In other cases, the solution to a problem makes its appearance indirectly in symbolic guise. Elias Howe, the nineteenth-century inventor of the locked-stitch sewing machine, was baffled by the problem of how to thread the machine's needle until he dreamed of being captured by cannibals who menaced him with spears and prepared to cook him. Peering out from the cauldron in which he was about to be stewed, Howe observed that the heads of his captors' spears had eye-shaped holes in them. He awoke with the realization that the needle used in the sewing machine needed to be threaded near its *point*. The German chemist Friedrich August Kekulé discovered the tetravalence of carbon during a dreamlike experience on a London bus; and many years later he had a dream of whirling snakes that allowed him to unlock the ring structure of the benzene molecule. Awakened by her cat in the middle of the night, evolutionary biologist Marjorie Profet dreamed "I had a vision of a cartoon from grade school. The film's little images showed ovaries, the uterus. But there were all these tiny black triangles with pointy tips embedded in the uterus and they were coming out with the flow." Profet interpreted the triangles as germs, which led to her influential hypothesis that the function of menstruation is to get rid of pathogenic microorganisms that have entered the uterus on the backs of visiting sperm.[11]

Psychological Biases and Defense Mechanisms

Psychologists have identified quite a few specific forms of self-deceptive cognitive bias, widespread tendencies to distort perceptions, inferences, and memories. Psychoanalysts, too, have presented a set of psychological "defense mechanisms" that might equally well be called "mechanisms of self-deception." I have itemized below a selection of the main forms of self-deception mentioned in the psychological and psychoanalytic literature.[1]

Psychological Biases

1. *Self-serving bias:* The tendency to take the credit for success, and blame external factors for failure.

2. *Self-centered bias:* The tendency for an individual contributor to take a disproportionate amount of credit for the outcome of a collective effort.

3. *Egocentricity bias:* The tendency to exaggerate the importance of one's role in past events.

4. *False consensus effect:* The tendency to believe that most people share one's opinions and values.

5. *Assumption of uniqueness:* The tendency to overestimate one's uniqueness.

6. *Illusion of control:* The tendency to exaggerate the degree of one's control over external events.

7. *Hindsight bias:* The tendency to retrospectively overestimate the probability of past events occurring.

8. *Self-righteous bias:* The tendency to regard oneself as having higher moral standards or greater moral consistency than others have.

9. *In-group/out-group bias:* The tendency to view members of the group to which one belongs in a more positive light than members of groups of which one is not a member. Out-group members are seen as less worthy, more responsible for their misfortunes, less responsible for their successes, and conforming more to stereotypes than in-group members.

10. *Base-rate fallacy:* The tendency to neglect population characteristics and prior probabilities when making probabilistic inferences.

11. *Conjunction fallacy:* The tendency to regard the conjunction of two events as more probable than either of them occurring singly.

Defense Mechanisms

1. *Repression:* Motivated amnesia.

2. *Disavowal (or denial):* Disbelieving a true memory or perception.

3. *Projection:* Misattributing some aspect of oneself to someone else.

4. *Introjection (or internalization, identification):* Misattributing an aspect of someone else to oneself.

5. *Displacement:* Redirecting emotions or attitudes from their proper object to a proxy.

6. *Retroflection:* Redirecting emotions or attitudes from their proper object to oneself.

7. *Reaction-formation:* Representing an attitude or emotion as its opposite.

8. *Negation:* Falsely believing that it is not the case that one has a particular attitude.

9. *Isolation:* Stripping affect from thought.

10. *Rationalization:* Attributing mental states to false reasons.

11. *Acting out:* Precipitately acting to preempt conscious awareness.

Notes

Preface

1. Alexander, R. D. "The search for a general theory of behavior." *Behavioral Science* 10 (1975): 96.
2. James, W. *The Principles of Psychology*, vol. 1. (New York: Dover, 1950), 163.
3. Barash, D. P. *The Whisperings Within*. Second edition. (New York: Harper & Row, 1982), 211.
4. Krebs, J. R., and Dawkins, R. "Animal signals: Mind-reading and manipulation." In eds. J. R. Krebs and N. B. Davies *Behavioural Ecology: An Evolutionary Approach*. (Sunderland, MA: Sinauer Associates, 1984).
5. Wittgenstein, L. *Culture and Value*. Ed. G. H. von Wright., trans. Peter Winch (Oxford: Basil Blackwell, 1978).

1. Natural-Born Liars

1. Byrne, R. W., and Whiten, A. "The thinking primate's guide to deception." *New Scientist*, Dec. 3, 1987, 54–57. Byrne, R. W. *The Thinking Ape: Evolutionary Origins of Intelligence* (New York: Oxford University Press, 1995).
2. Wilson, E. O. *Sociobiology: The New Synthesis* (Cambridge, MA: Harvard University Press, 1975).
3. Segerstråle, U. *Defenders of the Truth: The Battle for Science in the Sociobiology Debate and Beyond* (Oxford: Oxford University Press, 2000).
4. Goleman, D. *Vital Lies, Simple Truths: The Psychology of Self-Deception* (London: Bloomsbury, 1985), 15.
5. Hamilton, W. D. *The Narrow Roads of Gene-Land, Vol. 1: Evolution of Social Behavior* (New York: W. H. Freeman & Co., 1996), 14.
6. Bok, S. *Lying: Moral Choice in Public and Private Life* (New York: Pantheon, 1999).
7. Twain, M. "My First Lie and How I Got Out of It." In *The Man That Corrupted Hadleyburg and Other Stones and Sketches* (London: Oxford University Press, 1996).

8. Here is Marc Hauser's more technical specification of functional deception, from Hauser, M. "Minding the behavior of deception." In eds. A. Whiten and R. W. Byrne *Machiavellian Intelligence II: Extensions and Evaluations* (Cambridge: Cambridge University Press, 1997), 116:

> *For an animal's behavior to be considered functionally deceptive, the following conditions must hold: (a) There must exist a context, C, in which individuals typically (high probability) produce a signal, S_c, which causes other group members to respond with behavior, B; (b) Occasionally, individuals produce S in a different context $C'(S_c)$ which causes other group members to respond with behavior B and consequently, enables the signaller to experience a relative increase in fitness (active falsification); (c) Occasionally, individuals fail to produce S in context C, consequently enabling them to experience a relative increase in fitness (withholding information); (d) The fitness increase comes from the fact that the functionally deceptive act allows deceivers to gain some benefit whereas those who are deceived obtain some cost.*

9. O'Connell, S. *Mind reading: How We Learn to Love and Lie* (London: Arrow, 1998).

10. Carey, B. "A highly inflated version of reality: researchers challenge notions about what drives the chronic liar." *The Los Angeles Times*, March 3, 2003.

11. Alexander, R. D. *The Biology of Moral Systems* (New York: Aldine de Gruyter, 1987), 198.

12. Power, C., and Watts, I. "Female strategies and collective behaviour: the archaeology of earliest *Homo sapiens sapiens*." In J. Steele and S. Zaran (eds.) *The Archaeology of Human Ancestry: Power, Sex and Tradition* (London: Routledge, 1996), 306–30. There is another intriguing possibility about the origins of rouge. It is widely believed that the biological suppression of signs of estrus was to the advantage of hominid females because it contributed to the development of long-term pair bonding with males. In species in which estrus is not concealed, such as the chimpanzees, males are only interested in females when they are "in heat." It may be that because hominid males could not determine just when females were ovulating, they had to maintain an interest in them all the time. However, subtle changes announce ovulation in women. One of these is a slight "flush," or reddening, of the skin. If it was advantageous for hominid females to suppress the signs of ovulation but costly to males, we would expect that this would act as a selection pressure for males to evolve the ability to detect these subtle signs. The early use of ochre may then have been selected to mimic the physiological sign of estrus and thus prevent males from recognizing the true time of ovulation.

13. Angeloglou, M. *A History of Make-up* (New York: Macmillan, 1970). Many of these techniques create a misleading hormonal picture. Lip fullness, large eyes, the absence of skin blemishes, hair quality, and a lack of body hair indicate high estrogen levels and hence fertility. The use of makeup to lighten skin may imitate the faint lightening of a woman's skin as she enters estrus. For a fascinating account of the history of hair coloring, see Pitman, J. *On Blondes* (London: Bloomsbury, 2003).

14. Antiphanes' "Thorikan Villagers, or Digging Through," cited in Angeloglou, M. *A History of Make-up* (New York: Macmillan, 1970). There is evidence to suggest that olfactory deception does not necessarily involve replacing one's bodily scent with a different one. Research suggests that women often gravitate to fragrances that *amplify* rather than suppress their natural body odor, apparently because this advertises information about their immunogenetics to prospective mates. Milinski, M., and Wedekind, C. "Evidence for MHC-correlated perfume preferences in humans." *Behavioral Ecology* 12(2) (2001): 140–149.

15. Anderson, M. "Cultural concatenation of deceit and secrecy." In R. W. Mitchell and N. S. Thompson (eds.) *Deception: Perspectives on Human and Nonhuman Deceit* (Albany: State University of New York Press), 333.

16. O'Connell, *Mind reading: How We Learn to Love and Lie* (London: Arrow, 1998).

17. Knox, D., Schacht, C., Holt, J., et al. "Sexual lies among university students." *College Student Journal* 27 (1993): 269–272. See also Cochran, S. D., and Mays, V. M. "Sex, Lies and HIV." *New England Journal of Medicine*, 322 (1990): 774–775.

18. Underwood, J. "Truth, lies and resumes." *The Birmingham News*, August 22, 1993: D1, D10.

19. Alexander, M. G., and Fisher, T. D. "Truth and consequences: Using the bogus pipeline to examine sex differences in self-reported sexuality." *The Journal of Sex Research* 40 (2003): 27–35.

20. Goffman, E. *The Presentation of the Self in Everyday Life* (Harmondsworth: Penguin, 1969).

21. Reibstein, J., and Richards, M. *Sexual Arrangements: Marriage, Monogamy and Affairs* (London: Mandarin, 1993).

22. Blumstein, P., and Schwartz, P. *American Couples* (New York: McGraw Hill, 1983).

23. Cuthbert, S. A., and McDonough, J. J. "Trusting relationships, empowerment, and the conditions that produce truth-telling." In Massarik, F. (ed.) *Advances in Organization Development*, Vol. 2. (Norwood, NJ: Ablex Publishers, 1992). Jackall, R. "Structural invitations to deceit: some reflections on bureaucracy and morality." *Berkshire Review* 15 (1980): 49–61. Blumberg, P. *The Predatory Society: Deception in the American Marketplace* (New York: Oxford University Press, 1989). Warner, F. "What happened to the truth?" *Adweek's Marketing Week*, October 28, 1991: 3–4.

24. Novack, D. H., Detering, B. S., Farrow, L., et al. "Physicians' attitudes towards using deception to resolve difficult ethical problems." *Journal of the American Medical Association* 261 (1989): 2980–2985.

25. Sun Tzu. *The Art of War* (New York: Delacorte, 1983), 11.

26. Sun Tzu. *The Art of War* (New York: Delacorte, 1983), 20.

27. Researchers at the RAND Corporation, which advises the United States government, have produced a detailed report on how the military might benefit from the study of methods of deception used by other species. Gewehr, S., and Glenn, R. W. *Unweaving the Web: Deception and Adaptation in Future Urban Warfare* (Santa Monica, CA: The Rand Corporation).

28. Cited in Bourke, J. *An Intimate History of Killing* (New York: Basic Books, 1999), 101.

29. Bourke, J. *An Intimate History of Killing* (New York: Basic Books, 1999), 121.

30. Pattern, F., with Sugar, B. R. *Inside Boxing* (Chicago: Contemporary Books, 1974).

31. Westfall, R. S. "Newton and the fudge factor." *Science* 179 (1973): 751–758.

32. Broad, W., and Wade, N. *Betrayers of the Truth: Fraud and Deceit in Science* (Oxford: Oxford University Press, 1982).

33. Fisher, R. A. "Has Mendel's work been rediscovered?" *Annals of Science* 1 (1936): 115–137.

34. Cioffi, F. *Freud and the Question of Pseudoscience* (Chicago: Open Court, 1998). F. C. Crews (ed.) *Unauthorized Freud: Doubters Confront a Legend* (Harmondsworth: Penguin, 1998). Esterson, A. *Seductive Mirage: An Exploration of the Work of Sigmund Freud* (Chicago: Open Court, 1993).

35. Sartre, J-P. *Being and Nothingness: An Essay on Phenomenological Ontology*, trans. H. Barnes (New York: New York Philosophical Library, 1943). Gergen, K. "The ethnopsychology of self-deception." In M. Martin (ed.) *Self-Deception and Self-Understanding* (Topeka, KS: University of Kansas Press, 1985). Haight, M. *A Study of Self-Deception* (Brighton: Harvester Press, 1980). Kipp, D. "On self-deception." *Philosophical Quarterly* 30 (1980): 305–317.

36. Davidson, D. "Deception and division." In J. Elster (ed.) *The Multiple Self* (Cambridge: Cambridge University Press, 1986). Gardner, S. *Irrationality and the Philosophy of Psychoanalysis* (Cambridge: Cambridge University Press, 1993).

37. Sartre, J-P. *Being and Nothingness: An Essay on Phenomenological Ontology*, trans. H. Barnes (New York: New York Philosophical Library, 1943).

38. Paul Churchland, a contemporary philosopher who has little truck with Freud, includes hypnosis, dreams, mental illness, and organic disorders of the brain in his list of phenomena that folk-psychology is unable to explain. See Churchland, P. *A Neurocomputational Perspective* (Cambridge, MA: Bradford/MIT, 1989). For a detailed account of Freud's rejection of Cartesianism, see Smith, D. L. *Freud's Philosophy of the Unconscious* (Dordrecht, Netherlands: Kluwer Academic Publishers, 1999) and Schlepperman, H. *Dualism Schmualism* (New York: Alter Kacker Press, 1985).

39. As exemplified in George Miller's remark that "It is the *result* of thinking, not the process of thinking, that appears spontaneously in consciousness." Miller, G. A. *Psychology: The Science of Mental Life* (New York: Harper & Row, 1962), 56.

40. Nisbett, R. E., and Wilson, T. D. "Telling more than we can know: verbal reports on mental processes." *Psychological Review* 84(3) (1977): 231–259.

41. Gilovich, T. *How We Know What Isn't So* (New York: Macmillan, 1991).

42. Adams, H. E., Wright, Jr., L. W., and Lohr, B. A. "Is homophobia associated with homosexual arousal?" *Journal of Abnormal Psychology* 105 (1996): 440–445.

43. Latané, B., and Darley, J. M. *The Unresponsive Bystander: Why Doesn't He Help?* (New York: Appleton-Century-Crofts, 1970), 124.

44. Alloy, L. B., and Abramson, L. Y. "Judgement of contingency in depressed and nondepressed students: sadder but wiser?" *Journal of Experimental Psychology: General* 108(4) (1979): 441–485. Alloy, L. B., and Abramson, L. Y. "Learned helplessness, depression and the illusion of control." *Journal of Personality and Social Psychology* 42 (1982): 1114–1126.

45. Lewinsohn, P. M., Mischel, W., Chaplin, W., et al. "Social competence and depression: The role of illusionary self-perceptions." *Journal of Abnormal Psychology* 89 (1980): 203–212. Alloy, L. B., and Abramson, L. Y. "Depression and pessimism for the future: Biased use of statistically relevant information in predictions of self versus others." *Journal of Personality and Social Psychology* 41 (1987): 1129–1140.

46. Lane, R. D., Merikangas, K. R., Schwartz, G. E., et al. "Inverse relationship between defensiveness and lifetime prevalence of psychiatric disorder." *American Journal of Psychiatry* 147 (1990): 573–578. Sackheim, H. A., and Gur, R. C. "Self-deception, other deception, and self-reported psychopathology." *Journal of Consulting Clinical Psychology* 47 (1979): 213–215. Sackheim, H. A., and Wegner, A. Z. "Attributional patterns in depression and euthymia." *Archives of General Psychiatry* 43 (1986): 553–560. Taylor, S. E., and Brown, J. D. "Illusion and well-being: a social psychological perspective on mental health." *Psychological Bulletin* 103 (1988): 193–210.

47. Nyberg, D. *The Varnished Truth: Truth Telling and Deceiving in Ordinary Life* (Chicago: University of Chicago Press, 1993), 85.

48. Alexander, R. D. "The search for a general theory of behavior." *Behavioral Science* 20(1975):77–100.

2. Manipulators and Mind Readers

1. Zimmer, C. *Parasite Rex* (New York: Touchstone, 2000). Krause, J., Ruxton, G. D., and Godin, J. "Distribution of *Crassiphiala bulgoglossa*, a parasitic worm in shoaling fish." *Journal of Animal Ecology* 68 (1999): 27–33.

2. Zimmer, C. *Parasite Rex* (New York: Touchstone, 2000).

3. Maynard-Smith, J., and Harper, D. C. G. "Animal signals: models and terminology." *Journal of Theoretical Biology* 177 (1995): 305–311.

4. Trivers, R. L. *Social Evolution* (Menlo Park, CA: Benjamin Cummings, 1985), 401–402.

5. This comparison is intended quite seriously. John Krebs and Richard Dawkins note that "A man may be sexually aroused by a picture of a naked woman. A Martian ethologist, observing this, might regard the picture as 'mimicking' the real thing, and assume that the man was 'fooled' into 'thinking' it was a real woman." Krebs, J. R., and Dawkins, R. "Animal signals: mind-reading and manipulation." In J. R. Krebs and N. B. Davies, (eds.) *Behavioural Ecology: An Evolutionary Approach* (Sunderland, MA: Sinauer Associates, 1984), 385.

6. Krebs, J. R., and Dawkins, R. "Animal signals: mind-reading and manipulation." In J. R. Krebs, and N. B. Davies (eds.) *Behavioural Ecology: An Evolutionary Approach* (Sunderland, MA: Sinauer Associates, 1984).

7. Plotkin, H. *The Nature of Knowledge: Concerning Adaptations, Instinct and the Evolution of Knowledge* (London: Penguin, 1994). This view of knowledge is known as "evolutionary epistemology." See Campbell, D. T. "Blind variation and selective retention in creative thought as in other knowledge processes." *Psychological Review* 67 (1960): 380–400.

8. Krebs, J. R., and Dawkins, R. "Animal signals: mind-reading and manipulation." In J. R. Krebs, and N. B. Davies (eds.) *Behavioural Ecology: An Evolutionary Approach* (Sunderland, MA: Sinauer Associates, 1984), 386.

9. McFarland, D. "Camouflage." In D. McFarland (ed.) *The Oxford Companion to Animal Behaviour* (Oxford: Oxford University Press, 1987), 53–55.

10. Brown, L., and Amadon, E. *Eagles, Hawks and Falcons of the World* (Feltham: Country Life Books, 1968). Mueller, H. C. "Zone-tailed hawk and turkey vulture: mimicry or aerodynamics?" *Condor* 74 (1972): 221–222. Mueller, H. C. "Reaction of quail to flying vultures." *Condor* 78 (1976): 120–121.

11. Not all forms of mimicry are relevant to the topic of deception. Müllerian mimicry was discovered by the German zoologist Fritz Müller. Here's how it works. Imagine that you are a poisonous insect approached by an inexperienced predator who has not yet learned to avoid critters with your pattern of warning colors. Let's say that the predator eats you, gets violently ill, and subsequently avoids your species. Well, that's fine for your friends and relations, but it doesn't do you any good! Müllerian mimicry, the tendency of poisonous species to resemble one another, is a variation on the "safety in numbers" theme. If a number of poisonous species look alike, a predator that learns to avoid one will also avoid all the rest. There is an obvious problem here. If the prey is *fatal* to the predator, the predator never learns to avoid it, and thus there is no point to innocuous species mimicking extremely toxic ones. In Mertensian mimicry, both deadly and innocuous species mimic a mildly noxious species that predators learn, from bitter experience, to avoid.

12. Huey, R. B., and Pianka E. R. "Natural selection for juvenile lizards mimicking noxious beetles." *Science* 195 (1977): 201–203.

13. Platt, D. R. "Natural history of the hognose snakes Heterodon platyrhinos and Heterodon nasicus." *University of Kansas, Museum of Natural History* 18(4) (1969): 253–420.

14. Norman, M., Finn, J., and Tregenza, T. "Dynamic mimicry in an Indo-Malayan octopus." *The Proceedings of the Royal Society, London* 268 (2001): 478:1755.

15. Pietsch, T. W., and Grobecker, D. B. *Frogfishes of the World: Systematics, Zoogeography, and Behavioural Ecology* (Palo Alto: Stanford University Press, 1987).

16. Lloyd, J. E. "Mimicry in the Sexual Signals of Fireflies." *Scientific American* 245 (July 1981): 139–145. Eisner, T., et al. "Firefly 'femmes fatales' acquire defensive steroid (lucibufagins) from their firefly prey." *Proceedings of the National Academy of Science* 94 (1997): 9723–9728.

17. Eberhard, W. G. "Aggressive chemical mimicry by a bolas spider." *Science* 198 (4322) (1977): 1173–1175. Gemeno, C., Yeargan, K. V., and Haynes, K. F. "Aggressive chemical mimicry by the bolas spider *Mastophora hutchinsoni*: Identification and quantification of a major prey's sex pheromone components in the spider's volatile emissions." *Journal of Chemical Ecology* 26 (2000): 1235–1243. Haynes, K. F., and Yeargan, K. V. "Exploitation of intraspecific communication systems: illicit signalers and receivers." *Annals of the Entomological Society of America* 92 (1999): 960–970. Haynes, K. F., Yeargan, K. V., Millar, J. G., and Chastain, B. B. "Identification of sex pheromone of *Tetanolita mynesalis* (Lepidoptera: Noctuidae), a prey species of bolas spider *Mastophora hutchinsoni*." *Journal of Chemical Ecology* 22 (1996): 75–89.

18. Jackson, R. R. "*Portia* spider mistress of deception." *National Geographic*, 190(5) (1996): 104–115. Tarsitano, M. S., and Jackson R. R. "Jumping spiders make predatory detours requiring movement away from prey." *Behaviour* 131 (1994): 65–73. Jackson, R. R., et al. "Interpopulation variation in the risk-related decisions of *Portia labiata*, an araneophagic jumping spider (Araneae, Salticidae), during predatory sequences with spitting spiders." *Animal Cognition* 5 (2002): 215–223. Jackson, R. R., and Wilcox, R. S. "Spider-eating spiders." *American Scientist* 86 (1998): 350–357.

19. Holldobler, B. "Communication between ants and their guests." *Scientific American* 224 (1971): 86–93.

20. Vitt, L. J., Congdon, J. D., and Dickson, N. A. "Adaptive strategies and energetics of tail autotomy in lizards." *Ecology* 58 (1977): 326–337.

21. Hanlon R. T., and Messenger J. B. *Cephalopod Behavior* (Cambridge University Press, 1996).

22. Candolin, U. "Increased signaling effort when survival prospects decrease: male-male competition ensures honesty." *Animal Behaviour* 60 (2000): 417–422.

23. Norman, M. D., Finn, J., and Tregenza, T. "Female impersonation as an alternative reproductive strategy in giant cuttlefish." *Proceedings of the Royal Society, London* 266 (1999): 1347–1349.

24. Field, S. A., and Keller, M. A. "Alternative mating tactics and female mimicry as post-copulatory mate-guarding behaviour in the parasitic wasp *Cotesia rubecula*." *Animal Behavior* 46 (1993): 1183–1189.

25. Thornhill, R. "Adaptive female-mimicking behavior in a scorpionfly." *Science* 295 (1979): 412–414.

26. Mizutani, A., Chahl, J. S., and Srinivasan, M. V. "Motion camouflage in dragonflies." *Nature* 423 (2003): 604. Srinivasan, M. V., & Davey, M. "Strategies for active camouflage of motion." *Proceedings of the Royal Society, London* 259 (1995): 19–25.

27. Shine, R., et al. "Benefits of female mimicry in snakes." *Nature* 414 (2001): 267.

28. Fincke, O. M. "On the difficulty of detecting density-dependent selection on polymorphic females of the damselfly *Ischnura graellsii*: failure to reject the null." *Evolutionary Ecology* 8 (1994): 328–329. Fincke, O. M. "Female colour polymorphism in damselflies: failure to reject the null

hypothesis." *Animal Behaviour* 47 (1994) 1249–1266. Forbes, M.R.L. "Female morphs of the damselfly *Enallagma boreale Selys* (Odonata: Coenagrionidae): a benefit for androchromatypes." *Canadian Journal of Zoology* 69 (1991): 1969–1970. Forbes, M. R. L. "Tests of hypothesis for female-limited polymorphism in the damselfly, *Enallagma boreale Selys.*" *Animal Behaviour* 47 (1994): 724–726.

29. McGregor, P. K., and Peake, T. M. "Communication networks: social environments for receiving and signaling behaviour." *Acta Ethologica* 2 (2000): 71–81. Peake T. M., Terry, A. M. R., McGregor, P. K., & Dabelsteen, T. "Male great tits eavesdrop on simulated male-to-male vocal interactions." *Proceedings of the Royal Society, London* 268 (2001): 1183–1187. Naguib, M., and Todt, D. "Effects of dyadic vocal interactions on other conspecific receivers in nightingales." *Animal Behaviour* 54 (1997): 1535–1543. Oliveira, R. F., Lopes, M., Carneiro, L. A., and Canário, A. V. M. "Watching fights raises fish hormone levels." *Nature* 409 (2001): 475.

30. Blest, A. D., Collett, T. S., and Pye, J. D. "The generation of ultrasonic signals by a New World artiid moth." *Proceedings of the Royal Society, London* 158 (1963): 196–207. Fullard, J. H., Simmons, J. A., and Saillant, P. A. "Jamming bat echolocation: the dogbane tiger moth *Cycnia tenera* times its clicks to the terminal attack calls of the big brown bat *Eptesicus fuscus.*" *Journal of Experimental Biology* 194 (1994): 285–289. Faure, P. A., Fullard, J. H., and Dawson, J. W. "The gleaning attacks of the northern long-eared bat, *Myotis septentrionalis*, are relatively inaudible to moths." *Journal of Experimental Biology* 178 (1993): 173–189. Bogdanowicz, W., Fenton, M. B., and Daleszczyk, K. "The relationships between echolocation calls, morphology and diet in insectivorous bats." *Journal of Zoology* 247 (1999): 381–383. Fullard J. H., and Dawson J. W. "The echolocation calls of the spotted bat *Euderma maculatum* are relatively inaudible to moths." *Journal of Experimental Biology* 200 (1997): 129–137. Lacki M. J., and Ladeur K. M. "Seasonal use of lepidopteran prey by Rafinesque's big-eared bats (*Corynorhinus rafinesquii*)." *American Midland Naturalist* 145 (2001): 213–217. Fenton, M. B., and Fullard J. H. "The influence of moth hearing on bat echolocation strategies." *Journal of Comparative Physiology* 132 (1979): 77–86.

31. Humphries, D. A., and Driver, P. M. "Protean defense by prey animals." *Oecologia* 5 (1970): 285–302. Miller, G. F."Protean primates: the evolution of adaptive unpredictability in competition and courtship." In Whiten, A., and Byrne, R. W. (eds.), *Machiavellian Intelligence II: Extensions and Evaluations* (Cambridge: Cambridge University Press, 1997). Arnott, S. A., Neil, D. A., and Ansell, A. D. "Tail-flip mechanism and size-dependent kinematics of escape swimming in the brown shrimp, *Crangon crangon.*" *Journal of Experimental Biology* 201 (1998): 1771–1784. Garcia-Paris, M., & Deban, S. M. "A novel antipredator mechanism in salamanders: rolling escape in *Hydromantes platycephalus.*" *Journal of Herpetology* 29 (1995): 149–151. It may be that systematic unpredictability is a specialty of our own species and forms a basis for our specious but much-vaunted claim to possess freedom of the will. See Evans, D. "The arbitrary ape." *New Scientist* 159 (2148) (1998): 32–35.

32. This taxonomy is loosely based on Robert W. Mitchell's hierarchy of deception. Mitchell, R. W. "A framework for discussing deception." In. Mitchell, R. W., and Thompson, N. S. (eds.) *Deception* (New York: State University of New York Press, 1986).

3. The Evolution of Machiavelli

1. Paley, W. *Natural Theology: or, Evidences of the Existence and Attributes of the Deity, Collected from the Appearances of Nature,* 12th Edition (New York: Classworks, 1986), 1–2.

2. Alternatively, God is sometimes pictured as a supreme version of another species (the Egyptian god Kephra, for example, was depicted as a dung beetle). However, even these deities, not to mention divine volcanoes and the like, possess human attributes. Pascal Boyer provides a powerful cognitive-evolutionary theory of why this should be. Boyer, P. *Religion Explained; The Evolutionary Origins of Religious Thought* (New York: Basic Books, 2001).

3. Trivers, R. L. *Social Evolution* (Menlo Park, CA.: Benjamin Cummings, 1985).

4. Etcoff, N. *The Survival of the Prettiest: The Science of Beauty* (New York: Anchor, 2000).

5. Dawkins, R. *The Blind Watchmaker* (New York: W. W. Norton, 1987).

6. Hrdy, S. B. *Mother Nature: Maternal Instincts and How They Shape the Human Species* (New York: Ballantine Books, 1999), 427.

7. Solomon, N., and French, J. (eds.) *Cooperative Breeding in Mammals* (Cambridge: Cambridge University Press, 1996). Rood, J. P. "Group size, survival, reproduction and routes to breeding in dwarf mongooses." *Animal Behavior* 39 (1990): 556–572. Sherman, P. "Nepotism and the evolution of alarm calls." *Science* 197 (1977): 1246–1253. For an alternative account of the significance of alarm calls see Zahavi, A., and Zahavi, A. *The Handicap Principle* (Oxford: Oxford U0niversity Press, 1997).

8. Wilson, E. O. *The Insect Societies* (Cambridge, MA: Harvard/Belknap, 1971).

9. Orthodox evolutionary theory states that group selection occurs in special circumstances and that the effect is usually a weak one. However, some biologists are prepared to give group selection wider scope and significance. See, for example, Wilson, D. S. "Introducing group selection into the adaptationist program: a case study involving human decision making." In Simpson, J. and Kendricks, J. (eds.) *Evolutionary Social Psychology* (Mawah, NJ: Lawrence Erlbaum, 1997). Wilson, D. S. "Human groups as units of selection." *Science* 276 (June 27, 1997): 1816–1817. Wilson, D. S., and Sober, E. "Reintroducing group selection to the human behavioral sciences." *Behavioral and Brain Sciences* 17 (December 1994): 585–654. Bloom, H. *Global Brain: The Evolution of Mass Mind* (New York: John Wiley & Sons, 2000). Keller, L. *Levels of Selection in Evolution* (Princeton, NJ: Princeton University Press, 1999). Sober, E., and Wilson, D. S. *Unto Others: The Evolution and Psychology of Unselfish Behavior* (Cambridge, MA: Harvard University Press,

1999). The *locus classicus* for modern group selectionism is Wynne-Edwards, V. C. *Animal Dispersion in Relation to Social Behavior* (London: Oliver & Boyd, 1962). The classic critique of group selectionism is Williams, G. C. *Adaptation and Natural Selection* (Princeton, NJ: Princeton University Press, 1966).

10. Hamilton, W. D. *The Narrow Roads of Gene-Land, Vol. 1: The Evolution of Social Behavior* (New York: W. H. Freeman, 1996), 25.

11. See Hamilton, W. D. "The evolution of social behavior." *Journal of Theoretical Biology* 7 (1964): 1–52, and Hamilton, W. D. "The evolution of altruistic behavior." *The American Naturalist* 97 (1963): 354–356. The eusociality of the hymenoptera has a special genetic underpinning. Social insects are *haplodiploid*. Females are diploid; that is, they possess one set of chromosomes from each parent. Males, however, are haploid—they are born from unfertilized eggs and therefore have only a single set of chromosomes from their mother. As a result, the coefficient of relatedness between sisters is .75 whereas the coefficient of relatedness between a female and her offspring would be only .5. It therefore makes biological sense for female worker ants to help their queen make more sisters instead of creating offspring themselves. Darwin anticipated Hamilton's theory, suggesting that the trait of sterility might be selected for if it contributed to the reproductive success of close relatives. His analysis was hampered by his ignorance of the genetic mechanism of natural selection. Although Hamilton's equations provide an elegant account of ant and bee eusociality, the explanation of termite altruism is more obscure, as they are not haplodiploid and cannot reap the same genetic gains from eusociality as their hymenopteran cousins.

12. Trivers, R. L. "The evolution of reciprocal altruism." *Quarterly Review of Biology* 46 (1971): 35–57.

13. Trivers, R. L. "The evolution of reciprocal altruism." *Quarterly Review of Biology* 46 (1971): 35–57. Krebs, J. R., and Dawkins, R. "Animal signals: mind-reading and manipulation." In Krebs, J. R., and Davies, N. B. (eds.) *Behavioural Ecology: An Evolutionary Approach* (Sunderland, MA: Sinauer Associates, 1984). Protracted contact with close kin is important because there is evidence to suggest that reciprocal altruism developed out of kin altruism.

14. Wilkinson, S. G., "Reciprocal food sharing in the vampire bat." *Nature* 308 (1984): 181–184. For other examples of reciprocal altruism see Connor, R. C., and Norris, K. "Are dolphins reciprocal altruists?" *American Naturalist* 119 (1982): 358–374. Cheney, D. L. "The acquisition of rank and the development of reciprocal alliances among free-ranging immature baboons." *Behavioral Ecology and Sociobiology* 2 (1977): 303–318. Packer, C. "Reciprocal altruism in *Papio anubis*." *Nature* 265 (1979): 441–443. Seyfarth, R. M., and Cheney, D. L. "Grooming alliances and reciprocal altruism in vervet monkeys." *Nature* 308 (1984): 541–543.

15. Brosnan, S. F., and de Waal, F. B. M. "Monkeys reject unequal pay." *Nature* 425 (2003): 297–299.

16. Trivers, R. L. "The evolution of reciprocal altruism." *Quarterly Review of Biology* 46 (1971): 35–57.

17. Freud used the term "repression" in two different senses at different points in his career. In its generic sense, repression is equivalent to self-deception per se. However, Freud also used the term to designate one form of self-deception among many. See Appendix II for the latter.

18. Humphrey, N. "The social function of intellect." in Bateson, P.P.G., and Hinde, R. A. (eds.) *Growing Points in Ethology* (Cambridge: Cambridge University Press, 1976), 309. See also Chance, M.R.A., and Mead, A. P. "Social behavior and primate evolution." *Symposia for the Society of Experimental Biology, Evolution* 7 (1953): 395–439, and Jolley, A. "Lemur social behavior and primate intelligence." *Science* 153 (1966): 501–506. The idea that the human brain is a social tool, and that its evolved function is manipulating others, also appears in the work of Richard Alexander from around 1967. (Richard Alexander, personal communication)

19. For chimpanzee violence see Wrangham, R., and Peterson, D. *Demonic Males* (Boston: Houghton Mifflin, 1996). For archeological evidence of human cannibalism, see Diamond, J. M. "Talk of cannibalism." *Nature* 407 (2000): 25–26, and Marlar, R. A., et al. "Biochemical evidence of cannibalism at a prehistoric Puebloan site in southwestern Colorado." *Nature* 407 (2000): 74–78. For genetic evidence see Mead, S., et. al. "Balancing Selection at the prion protein gene consistent with prehistoric kurulike epidemics." *Science* 300 (5619) (2003): 640–643. The Latin quote is from Plautus's *Asinaria* (II, 4, 88).

20. Dunbar, R. *Grooming, Gossip and the Evolution of Language* (London: Faber and Faber, 1996), 65. The point was apparently originally made in Menzel, E. W., and Johnson, M. K. "Communication and cognitive organization in humans and other animals." *Annals of the New York Academy of Sciences* 280 (1976): 131–142.

21. See Gallistel, C. R. and Gelman, R. "Preverbal and verbal counting and computation." *Cognition* 44 (1992): 43-74. Hauser, M. D., MacNeilage, P., and Ware, M. "Numerical representations in primates." *Proceedings of the National Academy of Sciences* 93 (1996): 1514–1517. Dahaene, S. *The Number Sense: How the Mind Creates Mathematics* (New York: Oxford University Press, 1997). Buttterworth, B. *The Mathematical Brain* (New York: Papermac, 2000). Gallistel, C. R., and Gelman, R. "Nonverbal numerical cognition: from reals to integers." *Trends in Cognitive Sciences* 4(2) (2000): 59–65. Naccache, L., and Dehaene, S. "The priming method: imaging unconscious repetition priming reveals an abstract representation of number in the parietal lobes." *Cerebral Cortex* 11(10) (2001): 966–974. Hauser, M. D., Tsao, F., Garcia, P., and Spelke, E. "Evolutionary foundations of number: spontaneous representation of numerical magnitudes by cotton-top tamarins." *Proceedings of the Royal Society, London* 270 (1523) (2003): 1441–1446.

22. Steiner, G. *After Babel* (New York: Oxford University Press, 1998), 218.

23. Bateson, G. "Problems of Cetacian and other mammalian communications." In *Steps to an Ecology of Mind* (New York: Ballantine, 1972).

24. Shakespeare, W. *Macbeth*, 1, vii.

25. Ekman, P. *Telling Lies.* 2nd edition. (New York: W. W. Norton & Co., 1985). While the zygomatic smile is controlled by the anterior cingulate region of the brain, which empowers emotional expression, the Duchenne smile is produced by the higher cortex.

26. Shakespeare, W. *Macbeth*, 1, iv.

27. Sapir, E. "The unconscious patterning of behaviour in society." In *Selected Writings of Edward Sapir* (Berkeley, CA: University of California Press, 1986).

28. Ekman, P. *Telling Lies.* 2nd edition. (New York: W. W. Norton & Co., 1985). Understanding the importance of nonverbal emotional seepage has led to important advances in lie-detection technology. A team at Manchester Metropolitan University in the north of England has developed a lie-detection device called the "Silent Talker" that uses a computer linked to a camera to detect and analyze subtle facial movements. The Silent Talker boasts an 80 percent accuracy rate detecting deception, in contrast to the 60 percent of the conventional polygraph test. The use of thermal imaging techniques to detect blushing too subtle for the naked eye boasts a comparable rate of accuracy. "Truth machine means that liars must keep a straight face." *The Independent,* January 27, 2003. Pavlidis, J., Eberhardt, N. L., and Levine, J. A. "Seeing through the face of deception" *Nature* 415 (2002): 35.

29. Rohwer, S. "Status signaling in Harris Sparrows: some experiments in deception." *Behavior* 61 (1977): 107-129. See also Moeller, A. P. "Social control of deception among status signalling house sparrows *Passer domesticus.*" *Behavioral Ecology and Sociobiology* 20 (1987): 307–311.

30. Hauser, M. "Minding the behavior of deception." In Whiten, A., and Byrne, R. W. (eds.) *Machiavellian Intelligence II: Extensions and Evaluations* (Cambridge: Cambridge University Press, 1997), 69.

31. Ekman, P. *Telling Lies.* 2nd edition. (New York: W. W. Norton & Co., 1992).

32. Trivers, R. L. "Introduction." In Lockhard, J. S., and Paulhus, D. L. (eds.) *Self-Deception: An Adaptive Mechanism?* (Englewood Cliffs, NJ: Prentice-Hall, 1988).

33. "Interview with Robert L. Trivers." *Omni* (July 1985): 77–111.

34. Trivers, R. L. "Sociobiology and politics." In White E. (ed.) *Human Sociobiology and Politics* (Lexington, MA: Lexington Books, 1981).

35. Asch, S. E. *Social Psychology* (New York: Prentice-Hall, 1952).

36. Ichheiser, G. "Misrepresentations of personality in everyday life." *Character and Personality* 11 (1943): 145–146.

4. The Architecture of the Machiavellian Mind

1. Trivers, R. L. "Sociobiology and politics." In E. White (ed.), *Sociobiology and Human Politics* (Lexington, MA: Lexington Books, 1981), 35.

2. Poincaré, H. "Mathematical creation." In B. Ghislin (ed.) *The Creative Process: Reflections on Invention in the Arts and Sciences* (Berkeley: University of California Press, 1952).

3. Poincaré, H. "Mathematical creation." In B. Ghislin (ed.), *The Creative Process: Reflections on Invention in the Arts and Sciences* (Berkeley: University of California Press, 1952), 26.

4. Poincaré placed great emphasis on the role of the unconscious mind in scientific work. According to his biographer, Toulouse, he made a point of getting to bed early so that he had ample time to do mathematical research in his sleep! Toulouse, E. *Henri Poincaré* (Paris, 1910).

5. Kanigel, R. *The Man Who Knew Infinity: A Life of the Genius Ramanujan* (New York: Washington Square Press, 1991).

6. An amusing survey of psychological research bearing on intuition is supplied by D. G. Myers, in *Intuition: Its Powers and Perils* (New Haven, CT: Yale University Press, 2002). For neuroscientific research into the distinctive physiological signature of the so-called "Aha" experience see Rodriguez, E., George N., Lachaux, J. P., Martinerie, J., Renault, B., and Varela, F. "Perception's shadow: Long-distance synchronization in the human brain activity." *Nature* 397 (1999): 430–433, and Auble, P. M., Franks, J. J., and Soraci, S. A., Jr. "Effort toward comprehension: Elaboration or 'aha!'?" *Memory & Cognition* 7 (1979): 426–434.

7. Ghislin, B. *The Creative Process: Reflections on Invention in the Arts and Sciences* (Berkeley: University of California Press, 1952), 27.

8. This is not as far-fetched as it might sound at first. Arthur Reber and his coworkers have demonstrated experimentally that conscious effort impedes unconscious learning of tasks involving complex stimuli. See Reber, A. S. "Implicit learning of synthetic languages: The role of instructional set." *Journal of Experimental Psychology: Human Learning and Memory* 2 (1976): 88–94. Reber, A. S., Kassin, S. M., Lewis, S., and Cantor, G. W. "On the relationship between implicit learning of a complex rule structure." *Journal of Experimental Psychology: Human Learning and Memory* 6 (1980): 492–502.

9. Mavromatis, A. *Hypnogogia* (London: Routledge & Kegan Paul, 1987).

10. In Hadamard, J. *The Psychology of Invention in the Mathematical Field* (Princeton, NJ: Princeton University Press, 1996), 8.

11. Darwin, C. *The Autobiography of Charles Darwin 1809–1882* (New York: W. W. Norton & Company, 1993).

12. "The Proof," PBS, October 28, 1997.

13. Lennenberg, E. H. *Biological Foundations of Language* (Chichester: John Wiley & Sons, 1967).

14. Piaget, J. "The affective unconscious and the cognitive unconscious." *Journal of the American Psychoanalytic Association* 21 (1973), 249.

15. Reber, A. S. *Implicit Learning and Tacit Knowledge: An Essay on the Cognitive Unconscious* (Oxford: Oxford University Press, 1993), 9.

16. Helmholtz, H. von. *Treatise on Physiological Optics*, Vol. 3., in R. I. Watson (ed.). *Basic Writings in the History of Psychology* (New York: Oxford University Press, 1979), 126.

17. Accessible accounts of Libet's experiments and their interpretations can be found in Nørretranders, T. *The User Illusion: Cutting Consciousness Down to Size* (New York: Viking/Penguin, 1998), and Tallis, F. *Hidden Minds: A History of the Unconscious* (New York: Arcade, 2002).

18. Libet, B. "Unconscious cerebral initiative and the role of conscious will in voluntary action." *Behavioral and Brain Sciences* 8 (1985): 529–566. There is some evidence that although the unconscious decision to move a Δhand precedes the conscious decision, the decision about *which* hand to move comes after the conscious decision to move a hand. See Trevena, J. A., and Miller, J. "Cortical movement preparation before and after a conscious decision to move." *Consciousness & Cognition: An International* 11(2) (2002): 162–190. For an excellent discussion of the illusory nature of conscious will, see Wegner, D. M. *The Illusion of Conscious Will* (Cambridge, MA: MIT Press, 2002).

19. Libet, B., Alberts, W. W., Wright, E. W., Jr., Feinstein, B., and Pearl, B. "Subjective referral of the timing for a conscious sensory experience." *Brain* 102 (1979): 193–224.

20. Grimes, J. "On the failure to detect changes in scenes across saccades." In K. Akins (ed.) *Perception* (New York: Oxford University Press, 1996).

21. Cheery, E. C. "Some experiments on the recognition of speech, with one and two ears." *Journal of the Acoustical Society of America* 25 (1953): 975–979.

22. For the original formulation of the bottleneck, see Broadbent, D. E. *Perception and Communication* (London: Pergamon Press, 1958). The experimental literature on unconscious semantic analysis is extensive. For some classic studies, see Corteen, R. S., and Wood, B. "Autonomic responses to shock-associated words in an unattended channel." *Journal of Experimental Psychology* 94 (1972): 308–313. Corteen, R. S., and Dunn, D. "Shock associated words in a non-attended message: A test for momentary awareness." *Journal of Experimental Psychology* 102 (1974): 1143–1144. Forster, K. I., and Govier, E. "Discrimination without awareness?" *Quarterly Journal of Experimental Psychology* 30 (1978): 289–295. Von Wright, J. M., Anderson, K., and Stenman, U. "Generalisation of conditioned GSRs in dichotic listening." In P. M. A. Rabbitt and S. Dornic (eds.) *Attention and Performance V* (London: Academic Press, 1975). For a good review, see Dixon, N. F. *Subliminal Perception: The Nature of a Controversy* (New York: McGraw-Hill, 1971). The extent to which we are unconsciously sensitive to semantic information remains a matter of controversy. Responses to Kihlstrom's question, "Is the unconscious smart or dumb?" range across the whole spectrum from "very smart" to "very dumb." For an excellent critical discussion see Pashler, H. *The Psychology of Attention* (Cambridge, MA: MIT Press, 1998).

23. MacKay, D. "Aspects of a theory of attention, memory and comprehension." *Quarterly Journal of Experimental Psychology* 25 (1973): 22–40.

24. This effect, known as "priming," was actually first described by the eighteenth-century German philosopher Gottfried Wilhelm von Leibniz, who wrote that:

> *There are hundreds of indications leading us to conclude that at every moment there is in us an infinity of perceptions, unaccompanied by awareness or reflection; that is, of alterations in the soul itself, of which we are unaware because the impressions are either too minute or too numerous, or else too unvarying, so that they are not sufficiently*

distinctive on their own. But when they are combined with others they do nevertheless have their effect and make themselves felt, at least confusedly, within the whole.

Leibniz, G. W. *New Essays on Human Understanding.* Trans. and ed. P. Remnant and J. Bennett (Cambridge: Cambridge University Press, 1981), 53.

25. Wegner, D. M. *The Illusion of Conscious Will* (Cambridge, MA: MIT Press, 2002), 57.

26. Bargh, J. A., Chen, M., and Burroughs, L. "Automaticity of social behavior: direct effects of trait construct and stereotype activation on action." *Journal of Personality and Social Psychology* 71 (1996): 230–234. Zajonc, R. B. "Feeling and thinking: preferences need no inferences." *American Psychologist* 35 (1980): 151–175.

27. For some representative views on the function of consciousness, see Schiffrin, R. M., and Schneider, W. "Controlled and automatic human information processing, II: Perceptual learning, automatic attending, and a general theory." *Psychological Review* 84 (1977): 127–190. Posner, M. I., and Snyder, C. R. R. "Faciliation and inhibition in the processing of signals." In P.M.A. Rabbitt, and S. Dornick (eds.) *Attention and Performance, V* (New York: Academic Press, 1975). Mandler, G. *Cognitive Psychology: An Essay in Cognitive Science* (Hillsdale, NJ: Erlbaum, 1985). Underwood, G. "Memory systems and conscious processes." In G. Underwood, G., and R. Stevens (eds.) *Aspects of Consciousness* (New York: Academic Press, 1979). For an argument that none of these is satisfactory, see Velmans, M. "Is human information processing conscious?" *Behavioral and Brain Sciences* 14 (1991): 651–726.

28. Freud, S. "The origins of psychoanalysis." In *The Standard Edition of the Complete Psychological Works of Sigmund Freud,* vol. 1. (London: Hogarth Press and the Institute of Psychoanalysis, 1953–1964). For the psychoanalytic roots of backpropagation, see Werbos, P. J. *The Roots of Backpropagation: From Ordered Derivatives to Neural Networks and Political Forecasting* (Hoboken, NJ: Wiley-Interscience, 1994).

29. Freud, S. "The origins of psychoanalysis." In *The Standard Edition of the Complete Psychological Works of Sigmund Freud,* vol. 1. (London: Hogarth Press and the Institute of Psychoanalysis, 1953–1964), 308.

30. Smith, D. L. *Freud's Philosophy of the Unconscious, Studies in Cognitive Systems,* vol. 23 (Dordrecht and Boston: Kluwer Academic Publishers).

31. The pioneer psychologist Carl Lashley, who certainly ranks highly in the psychological pantheon, expressed more or less the same position in his provocative dictum that "No activity of mind is ever conscious." This is echoed by contemporary psycholinguist Ray Jackendoff's "thought is never consciousness."

32. Lashley, K. "Cerebral organization of behavior." In Solomon, H., Cob, S., and Penfield, W. (eds.) *Brain and Human Behavior* (Baltimore: Williams and Wilkins, 1956). Jackendoff, R. "How language helps us think." *Pragmatics and Cognition* 4 (1996): 1–34. Velmans, M. "Is human information processing conscious?" *Behavioral and Brain Sciences* 14 (1991): 651–726.

33. Tooby, J., and Cosmides, L. "Psychological foundations of culture." In Barkow, J. H., Cosmides, L., and Tooby, J. (eds.) *The Adapted Mind: Evolutionary Psychology and the Generation of Culture* (New York: Oxford University Press, 1992). Gigerenzer, G. "The modularity of social intelligence." In Whiten, A., and Byrne, R. W. (eds.) *Machiavellian Intelligence II: Extensions and Evaluations* (Cambridge: Cambridge University Press, 1997).

34. Samuels refers to these two approaches as "strong" and "weak" versions of the modularity thesis. Samuels, R. "Massively modular minds: evolutionary psychology and cognitive architecture." In Carruthers, P., and Chamberlain, A. (eds.), *Evolution and the Human Mind: Modularity, Language and Meta-Cognition* (Cambridge: Cambridge University Press, 2000).

35. Cosmides, L., and Tooby, J. "Cognitive adaptations for social exchange." In Barkow, J. H., Cosmides, L., and Tooby, J. (eds.) *The Adapted Mind: Evolutionary Psychology and the Generation of Culture* (New York: Oxford University Press, 1992).

36. Rosin, P., Haidt, J., and McCauley, C. R. "Disgust." In Lewis, M., and Haviland, J. M. (eds.) *Handbook of Emotions* (New York: Guilford, 1993). Boyer, P. *Religion Explained* (New York: Basic Books, 2001).

37. Selfridge, O. "Pandemonium: a paradigm for learning." In *Symposium on the Mechanization of Thought Processes* (London: HM Stationery Office, 1959). See also Dennett, D. C. *Consciousness Explained* (Boston: Little Brown, 1991), and Dennett, D. C., and Kinsbourne, K. "Time and the observer." *Behavioral and Brain Sciences* 15(2) (1992): 183–247.

38. Dennett, D. C. *Kinds of Minds: Towards an Understanding of Consciousness* (London: Weidenfeld & Nicholson, 1997), 155.

5. Social Poker

1. Trivers, R. L. "Introduction." In Dawkins, R. *The Selfish Gene* (New York: Oxford University Press, 1976).

2. Boulton, M. J., and Smith, P. K. "The social nature of play fighting and play chasing: mechanisms and strategies underlying cooperation and compromise." In Barkow, J. H., Cosmides, L., and Tooby, J. (eds.) *The Adapted Mind: Evolutionary Psychology and the Generation of Culture* (New York: Oxford University Press, 1992). Freyd, J. J. "Dynamic representations guiding adaptive behavior." In Macar, F., Pouthas, V., and Friedman, W. J. (eds.), *Time, Action and Cognition: Towards Bridging the Gap* (Dordrecht: Kluwer Academic Publishers, 1992). Also Premack, D. "The infant's theory of self-propelled objects." *Cognition* 36 (1990): 1–16.

3. Alexander, R. D. "The evolution of the human psyche." In Mellars, and Stringer, C. (eds.) *The Human Revolution* (Princeton: Princeton University Press, 1989).

4. Hayano, D. M. *Poker Faces: The Life and Work of Professional Card Players* (Berkeley: University of California Press, 1982), 66. For a fascinating discussion of the biological roots of this strategy, see Miller, G. "Protean primates: the evolution of adaptive unpredictability in competition and courtship." In

Whiten, A., and Byrne, R. W. (eds.), *Machiavellian Intelligence II: Extensions and Evaluations* (Cambridge: Cambridge University Press, 1990).

5. Bellin, A. *Poker Nation* (New York: HarperCollins, 2002).

6. Arthur S. Reber, personal communication.

7. Ford, C. V. *Lies! Lies! Lies! The Psychology of Deceit* (Washington, D.C.: American Psychiatric Press, 1996), 216.

8. Dimberg, U., Thunberg, M., and Elmehed, K. "Unconscious facial reactions to emotional facial expressions." *Psychological Science* 11(1) (2000): 86. Ekman, P. *Telling Lies*, 2nd edition (New York: W. W. Norton & Co., 1992).

9. DePaolo, B. M., and Pfeiffer, R. L. "On-the-job experience at detecting deception." *Journal of Applied Social Psychology* 16 (1986): 249–267. Vrij, A. "Credibility judgments of detectives: the impact on nonverbal behavior, social skills and physical characteristics on impression formation." *Journal of Social Psychology* 133 (1993): 601–610. Paul Ekman and Maureen O'Sullivan confirmed that police officers, and most other professional groups, are lousy at consciously spotting deception. Many, but by no means all, Secret Service agents were very good at it, with psychiatrists running a distant second. Ekman, P., and O'Sullivan, M. "Who can catch a liar?" *American Psychologist* 46 (1991): 913–920. Ekman, P. *Telling Lies: Clues to Deceit in the Marketplace Politics and Marriage* (New York: Norton, 1992).

10. Trivers, R. L. "Sociobiology and politics." In White, E. (ed.) *Sociobiology and Human Politics* (Lexington, MA: Lexington Books, 1981), 35.

11. Mithen, S. *The Prehistory of the Mind* (London: Thames & Hudson, 1996).

12. Alexander, R. D. "The evolution of the human psyche." In Mellars, P. and Stringer, C. (eds.) *The Human Revolution* (Princeton: Princeton University Press, 1989).

13. Freud, S. "Fragment of an analysis of a case of hysteria." *Standard Edition of the Complete Psychological Works of Sigmund Freud*, Vol. 7. Trans. James Strachey (London: Hogarth Press and the Institute of Psychoanalysis 1953–1964), 55–56.

14. Freud, S. "The question of lay analysis." *Standard Edition of the Complete Psychological Works of Sigmund Freud*, Vol. 20. Trans. James Strachey (London: Hogarth Press and the Institute of Psychoanalysis, 1953–1964).

15. Kitcher, P. *Freud's Dream: A Complete Interdisciplinary Science of Mind* (Cambridge, MA: MIT Press). Smith, D. L. *Psychoanalysis in Focus*. (London: Sage, 2003).

16. "Interview with R. L. Trivers." *Omni* (July 1985), 111.

17. See the famous adaptation by Leda Cosmides of the Wason Selection Task (WST), a psychological experiment purporting to show that human beings are extraordinarily bad at understanding simple principles of deductive reasoning. Cosmides replaced the rather abstract structure of WST with a narrative with an identical logical form involving someone defaulting on a social rule, and found that subjects had no difficulty making the correct deductive inferences, apparently because when recast in this way the experiment activated a cognitive module specifically concerned with so-

cial reasoning. Cosmides, L. "The logic of social exchange: has natural selection shaped how humans reason? Studies with the Wason Selection Task." *Cognition* 31 (1989): 187–276. This interpretation of their results has not gone unchallenged. For a critique see Atran, S. "A cheater-detection module? Dubious interpretations of the Wason Selection Task and logic." *Evolution and Cognition* 7(2) (2001): 187–192. For the neuroscientific evidence, see Stone, V., Cosmides, L., Tooby, J., Kroll, N., and Knight, R. T. "Selective impairment of reasoning about social exchange in a patient with bilateral limbic system damage." *Proceedings of the National Academy of Sciences.* Published online, doi:10.1073/pnas. 122352699 (2002).

18. Freud, S. "The disposition to obsessional neurosis." *The Standard Edition of the Complete Psychological Works of Sigmund Freud,* Vol. 12 (London: Hogarth Press and the Institute of Psychoanalysis, 1953–1964), 320.

19. Freud, S. "Totem and taboo." *The Standard Edition of the Complete Psychological Works of Sigmund Freud,* Vol. 13 (London: Hogarth Press and the Institute of Psychoanalysis, 1953–1964), 139.

20. Myers, P. "Sándor Ferenczi and patients' perceptions of analysis." *British Journal of Psychotherapy* 13(1) (1996): 26–36. Smith, D. L. "Ferenczi, Langs and scientific reasoning: a reply to Martin Stanton." *British Journal of Psychotherapy* 149(3) (1998): 348–352. Smith, D. L. "The mirror-image of the present: Freud's first theory of retrogressive screen memories." *Psychoanalytische Perspectieven* 39 (2000): 7–28.

21. Ferenczi, S. "Confusion of tongues between adults and the child (the language of tenderness and of passion)." In Balint, M. (ed.), trans. M. Balint, M., and E. Mosbacher. *Final Contributions to the Problems and Methods of Psycho-Analysis* (London: Hogarth Press, 1955), 293, 296.

22. Ferenczi, S. "Confusion of tongues between adults and the child (the language of tenderness and of passion)." In Balint, M. (ed.), trans. M. Balint, M., and E. Mosbacher. *Final Contributions to the Problems and Methods of Psycho-Analysis* (London: Hogarth Press, 1955), 302.

23. Ferenczi, S. *The Clinical Diary of Sándor Ferenczi.* J. Dupont (ed.) trans. N. Jackson (Cambridge, MA: Harvard University Press, 1988), 1–2. It is worth noting that Freud, too, sometimes departed from the standard Freudian approach. Early in his career, he toyed witΔ221
h a theory of memory evocation very similar, if not identical, to Ferenczi's. See Smith, D. L. "The mirror-image of the present: Freud's first theory of retrogressive screen memories." *Psychoanalytische Perspectieven* 39 (2000): 7–28.

24. Ferenczi, S. "Notes and fragments." In Balint, M. (ed.), trans. M. Balint, M., and E. Mosbacher. *Final Contributions to the Problems and Methods of Psycho-Analysis* (London: Hogarth Press, 1955). Myers notes that the English translation of *ganz aktuellen* as "real-life" is inaccurate, and completely alters the meaning of the relevant passage. I have therefore used the literal translation "totally current" suggested in Myers, P. "Sándor Ferenczi and patients' perceptions of analysis." *British Journal of Psychotherapy* 13(1) (1996): 26–36.

6. Hot Gossip

1. This is a very simplified statement of the philosophical semantics of Willard Van Orman Quine. Quine, W.V.O. *Pursuit of Truth* (Cambridge, MA: Harvard University Press, 1990).

2. This marvelous phrase is actually the title of an equally marvelous paper by the philosopher Daniel C. Dennett. It was just too good not to borrow. Dennett, D. C. "A cure for the common code?" In *Brainstorms: Philosophical Essays on Mind and Psychology* (Brighton: Harvester Press, 1986).

3. Taylor, G. "Gossip as moral talk." In Goodman, R. F., and Ben Ze'ev, A. (eds.) *Good Gossip* (Lawrence, KS: University Press of Kansas, 1994), 34.

4. Abelson, R. P. "Computer simulation of 'hot cognitions'." In Tomkins, S., and Messick, S. (eds.). *Computer Simulation and Personality: Frontier of Psychological Theory* (New York: John Wiley & Sons, 1963).

5. Hume, D. *Treatise of Human Nature.* Selby-Bigge, L. A. (ed.) (Oxford: Oxford University Press, 1985), 416.

6. There is a sizable empirical literature on automaticity, most of which concerns relatively "stupid" routine processes and actions. See Uleman, J. S., and Bargh, J. A. (eds.) *Unintended Thought* (New York: Guilford Press, 1989).

7. Sterling, R. B. "Some psychological mechanisms operative in gossip." *Social Forces* 34 (1956): 262–267. Emler, N. "Gossip, reputation and social adaptation." In Goodman, R. F., and Ben Ze'ev, A. (eds.) *Good Gossip* (Lawrence, KS: University Press of Kansas, 1994). Schein, S. "Used and abused: gossip in medieval society." In Goodman, R. F., and Ben Ze'ev, A. (eds.) *Good Gossip* (Lawrence, KS: University Press of Kansas, 1994). Forrester, J. *The Polite Philosopher: Or, an Essay on that Art which Makes a Man Happy in Himself, and Agreeable to Others,* 2nd Edition (London: J. Wilson), 25–26.

8. Garrick, D. "Prologue." In *The Plays and Poems of Richard Brinsley Sheridan,* vol. 3. R. Compton Rhodes (ed.) (New York: Macmillan, 1929), 23.

9. Emler, N. "Gossip, reputation and social adaptation." In Goodman, R. F., and Ben Ze'ev, A. (eds.) *Good Gossip* (Lawrence, KS: University Press of Kansas, 1994). For additional sources on the universality of gossip, see Barkow, J. H. "Beneath new culture is old psychology: gossip and social stratification." In Barkow, J. H., Cosmides, L., and Tooby, J. (eds.) *The Adapted Mind: Evolutionary Psychology and the Generation of Culture* (New York: Oxford University Press, 1992). Cox, B. A. "What is Hopi gossip about? Information management and Hopi factions." *Man* 5 (1970): 88–98. Hess, N. C., and Hagen, E. H. "Informational warfare." CogPrints ID cog00002112. Lee, R. B. "Eating Christmas in the Kalahari." In Spradley, J. B., and McCurdy, D.H. (eds.) *Conformity and Conflict: Readings in Cultural Anthropology,* 7th Edition (Glenville, IL: Scott, 1990). McAndrew, F. T., and Melenkovic, M. A. "Of tabloids and family secrets: the evolutionary psychology of gossip." *Journal of Applied Social Psychology* 32 (2002): 1–20. McPherson, N. M. "A question of morality: sorcery and concepts of deviance among the Kabana, West New Britain." *Anthropologica* 33 (1991): 127–143. Paine, R. "What is gossip?" *Man* 2 (1967): 278–285.

10. Donald, M. *Origins of the Modern Mind: Three Stages in the Evolution of Culture and Cognition* (Cambridge, MA: Harvard University Press, 1991), 257.

11. Cheney, D. L., and Seyfarth, R. M. *How Monkeys See the World: Inside the Mind of Another Species* (Chicago: University of Chicago Press, 1990).

12. Aiello, L. C., and Dunbar, R. I. M. "Neocortex size, group size and the evolution of language." *Current Anthropology* 34 (1993): 184–193.

13. Aiello, L. C., and Dunbar, R. I. M. "Neocortex size, group size and the evolution of language." *Current Anthropology* 34 (1993): 184–193.

14. Trivers, R. L. "Parental investment and sexual selection." In B. Campbell (ed.) *Sexual Selection and the Descent of Man: 1871–1971* (Chicago: Aldine, 1972).

15. See Hill, K., & Kaplan, H. "Tradeoffs in male and female reproductive strategies among the Ashe: parts I and II." In Betzig, L., Borgerhoff Mulder, M., and Turk, B. (eds.) *Human Reproductive Behavior: A Darwinian Perspective* (Cambridge: Cambridge University Press, 1988). Key, C. A. and Aeillo, L. C. "The evolution of social organization." In Dunbar, R.I.M., Knight, C., and Power, C. (eds.) *The Evolution of Culture.* (Edinburgh: Edinburgh University Press, 1999). Power, C. "Secret language use at female initiation: bounding gossiping communities." In Knight, C., Studdert-Kennedy, M., and Hurford, J. R. (eds.) *The Evolutionary Emergence of Language* (Cambridge: Cambridge University Press, 2000).

16. Buss, D. M. *The Evolution of Desire: Strategies of Human Mating* (New York: Basic Books, 1994). Hagen and Hess suggest that another selection pressure for the evolution of gossip among women may have been the presence of intense intra-group competition in consequence of patrilocality. See Hess, N. C. and Hagen, E. H. "Informational warfare." CogPrints ID cog00002112.

17. Dunbar, R.I.M. "Culture, honesty and the free-rider problem." In Dunbar, R.I.M, Knight, C., and Power, C. (eds.) *The Evolution of Culture* (Edinburgh: Edinburgh University Press, 1999). Emler, N. "Gossip, reputation and social adaptation." In Goodman, R. F., and Ben Ze'ev, A. (eds.) *Good Gossip.* (Lawrence, KS: University Press of Kansas, 1994). Gluckman, M. "Gossip and scandal." *Current Anthropology* 4 (1963): 307–316. Sabini, J., and Silver, M. *Moralities of Everyday Life* (Oxford: Oxford University Press, 1990).

18. Barthes, R. *Roland Barthes by Roland Barthes*, trans. R. Howard (New York: Hill & Wang, 1977), 169.

19. This terminology is from Hess, N. C., and Hagen, E. H. "Informational warfare." CogPrints ID cog00002112.

20. Spacks, P. M. *Gossip* (New York: Alfred A. Knopf, 1985), 4.

21. Power, C. "Old wives' tales: the gossip hypothesis and the reliability of cheap signals." In Knight, C., Studdert-Kennedy, M., and Hurford, J. R. (eds.) *The Evolutionary Emergence of Language* (Cambridge: Cambridge University Press, 2000).

22. For displaced reference, see Hockett, C. F. "The origin of speech." *Scientific American* 202(3) (1960): 89–96.

23. Spacks, P. M. *Gossip* (New York: Alfred A. Knopf, 1985), 15.

24. Barkow, J. H. "Beneath new culture is old psychology: gossip and social stratification." In Barkow, J. H., Cosmides, L., and Tooby, J. (eds.) *The*

Adapted Mind: Evolutionary Psychology and the Generation of Culture (New York: Oxford University Press, 1992).

25. Alexander, R. D. "The evolution of the human psyche." In Mellars, P., and Stringer, C. (eds.) *The Human Revolution* (Princeton: Princeton University Press, 1989), 475.

7. Machiavelli on the Couch

1. For a similar, biological account of resistance, see Slavin, M. O., and Kriegman, D. *The Adaptive Design of the Human Psyche: Psychoanalysis, Evolutionary Psychology and the Therapeutic Process* (New York: Guilford, 1992).

2. In an effort to prove that such powers really exist, he even engineered telepathic experiments involving himself, Freud, and Anna Freud, and organized a spiritualistic séance in Freud's home. Smith, D. L. "Freud and the occult." In Erwin, E. (ed.) *The Freud Encyclopedia* (New York: Routledge, 2002).

3. Langs is an extremely prolific writer. The best overall introduction to his work is probably Langs, R. J. *Science, Systems and Psychoanalysis* (London: Karnac, 1992). Other works include Langs, R. J. *The Bipersonal Field* (New York: Jason Aronson, 1976); Langs, R. J. *The Listening Process* (New York: Jason Aronson, 1978); Langs, R. J. *Technique in Transition* (New York: Jason Aronson, 1978); Langs, R. J. *The Therapeutic Environment* (New York: Jason Aronson, 1979); Langs, R. J. *Interactions: The Realm of Transference and Countertransference* (New York: Jason Aronson, 1980); Langs, R. J. *Resistances and Interventions: The Nature of Psychotherapeutic Work* (New York: Jason Aronson, 1981); Langs, R. J. *Madness and Cure* (Emerson, NJ: Newconcept Press, 1985); and Langs, R. J. *A Clinical Workbook for Psychotherapists* (London: Karnac, 1992). For a summary statement of Langs's views on the evolution of the mind, see Langs, R. J. "The evolution of the unconscious processing systems of the human mind." *Theoria et Histori Scientifiarum* 7(2) (2003): 77–86.

4. Kenrick, D. T., Sadalla, E. K., and Keefe, R. C. "Evolutionary cognitive psychology: the missing heart of modern cognitive science." In Barkow, J. H., Cosmides, L., and Tooby, J. (eds.) *The Adapted Mind: Evolutionary Psychology and the Generation of Culture* (New York: Oxford University Press, 1992).

5. Langs does not provide any principled elucidation of why this should be so. He does *attempt* to, but frames his explanation mainly in terms of the general significance of boundaries for living systems, and conflates natural laws with normative rules. Langs notes that although psychotherapy patients tend to uphold the ground rules unconsciously, they also tend to violate them, as do psychotherapists. He tries to solve this conundrum by proposing that the "ideal" rules also evoke a type of anxiety that he calls "existential death anxiety." This explanation seems completely ad hoc and begs more questions than it answers. There is no principled explanation why these particular rules should be beneficial, why they should produce anxiety, or why a hypothetical death anxiety is the best explanation for deviations from them. In the absence of any real evidence that honoring the ground rules has any psychotherapeutic effect whatsoever, this "explanation" amounts to little

more than piling one uncertainty on top of another. See Langs, R. *Ground Rules in Psychotherapy and Counselling* (London: Karnac, 1998).

6. Jacobs, M. (ed.) *In Search of Supervision* (Buckingham: Open University Press, 1996), 21–22.

7. Little, M. "Countertransference and the patient's response to it." *International Journal of Psycho-Analysis* 32 (1951), p. 37.

8. Brown, D. E. *Human Universals* (New York: McGraw-Hill, 1991). De Waal, F.B.M. *Good Natured: The Origins of Right and Wrong in Humans and Other Animals* (Cambridge, MA: Harvard University Press, 1996). Ridley, M. *The Origins of Virtue* (New York: Viking, 1996); Flack, J. C., and de Waal, F.B.M. "Any animal whatever: Darwinian building-blocks of morality in monkeys and apes." In Katz, L. D. (ed.) *Evolutionary Origins of Morality: Cross-Disciplinary Perspectives*, (Bowling Green, OH: Imprint Academic, 2000).

9. Alexander, R. D. *Darwinism and Human Affairs* (Seattle: University of Washington Press, 1979).

10. Alexander, R. D. *Darwinism and Human Affairs* (Seattle: University of Washington Press, 1979).

11. Alexander, R. D. *Darwinism and Human Affairs* (Seattle: University of Washington Press, 1979), 275.

8. Conspiratorial Whispers and Covert Operations

1. Krebs, J. R., and Dawkins, R. "Animal signals: mind-reading and manipulation." In Krebs, J. R., and Davies, N. B. (eds.) *Behavioural Ecology: An Evolutionary Approach* (Sunderland, MA: Sinauer Associates, 1984), 386–391.

2. I am deeply indebted to the work of Ruth Garrett Millikan for this analysis of psychological knowledge. Millikan, R. G. *White Queen Psychology and Other Essays for Alice* (Cambridge, MA: Bradford/MIT, 1993).

3. Nina Strohminger, personal communication.

4. Wrangham, R. "African apes: the significance of African apes for reconstructing social evolution." In Kinzey, W. G. (ed.) *The Evolution of Human Behavior: Primate Models* (Albany, NY: SUNY Press, 1987). Boehm, C. *Hierarchy in the Forest* (Cambridge, MA: Harvard University Press, 1999). Boehm, C. "Conflict and the evolution of social control." In Katz, L. D. (ed.) *Evolutionary Origins of Morality: Cross-Disciplinary Perspectives* (Bowling Green, OH: Imprint Academic, 2000), 79–101.

5. Boehm, C. *Hierarchy in the Forest: The Evolution of Egalitarian Behavior* (Cambridge, MA: Harvard University Press, 1999). Boehm, C. "Conflict and the evolution of social control." In Katz, L. D. (ed.) *Evolutionary Origins of Morality: Cross-Disciplinary Perspectives* (Bowling Green, OH: Imprint Academic, 2000), 79–101.

6. Pinker, S. *The Blank Slate: The Modern Denial of Human Nature* (New York: Viking, 2002).

7. For the term "counterdomination" see Erdal, D., and Whiten, A. "On human egalitarianism: an evolutionary product of Machiavellian status escalation?" *Current Anthropology* 35 (1994): 175–184. Erdal, D., and

Whiten, A. "Egalitarianism and Machiavellian intelligence in human evolution." In Mellars, P. A. and Gibson, K. R. (eds.) *Modeling the Early Human Mind* (Cambridge: McDonald Institute for Archaeological Research, 1996): 139–150. For the sociopolitical implications of agriculture see Rubin, P. H. *Darwinian Politics: The Evolutionary Origin of Freedom* (New Brunswick, NJ: Rutgers University Press, 2002).

    8. Boehm, C. "Conflict and the evolution of social control." In Katz, L. D. (ed.) *Evolutionary Origins of Morality: Cross-Disciplinary Perspectives* (Bowling Green, OH: Imprint Academic, 2000) 94.

    9. Haskell, R. E. "The matrix of group talk: An empirical method of analysis and validation." *Small Group Behavior* 13 (1982): 419–443.

    10. Haskell, R. E. *Between the Lines: Unconscious Meaning in Everyday Conversation* (New York: Plenum, 1999).

    11. Readers who want to find out more can consult the following: Haskell, R. E. "An analogic model of small group behavior." *International Journal of Group Psychotherapy* 28 (1978): 27–54; Haskell, R. E. "The matrix of group talk: an empirical method of analysis and validation." *Small Group Behavior* 13 (1982): 165–191; Haskell, R. E. "Cognitive structure and transformation: An empirical model of the psycholinguistic function of numbers in discourse." *Small Group Behavior* 14 (1983): 419–443; Haskell, R. E. "Social cognition and the non-conscious expression of racial ideology." *Imagination, Cognition and Personality* 6 (1986): 75–97; Haskell, R. E. "Analogical transforms: a cognitive theory of the origin and development of equivalence transformation, Part I, II." *LMS and Symbolic Activity* 4 (1989): 247–277; Haskell, R. E. "Cognitive operations and non-conscious processing in dreams and waking verbal reports." *Imagination, Cognition and Personality* 10 (1990–1991): 65–84; Haskell, R. E. "An analogical methodology for analysis and validation of anomalous cognitive and linguistic operations in small group (fantasy theme) reports." *Small Group Research* 22 (1991): 443–474; Haskell, R. E. *Between the Lines: Unconscious Meaning in Everyday Conversation* (New York: Plenum, 1999); Haskell, R. E. "Unconscious communication: communicative psychoanalysis and subliteral cognition." *Journal of the American Academy of Psychoanalysis* 27(3) (1999): 471–502; Haskell, R. E. *Deep Listening: Hidden Meanings in Everyday Conversation.* (Cambridge, MA: Perseus, 2001); Haskell, R. E. "The new cognitive unconscious: a logico-mathematic-structural (LMS) methodology and theoretical bases for subliteral $S_{ub}L_{it}$ Cognition and Language." *Evolution and Cognition* 8(2) (2002): 184–199; Haskell, R. E., and Badalamenti, A. F. "Algebraic structure of verbal narratives with dual meanings." *Mathematical and Computer Modeling* 37 (2003): 383–393; Haskell, R. E. "Is the unconscious 'smart' or 'dumb', and if it's smart, how smart is it?: one more time—with feeling." *Theoria et Historia Scientiarum* 7(2) (2003): 31–60; Haskell, R. E. "A logico-mathematic-structural methodology for the analysis and validation of sub-literal ($S_{ub}L_{it}$) language and cognition." *Journal of Mind and Behavior* 24(3/4) [in press]. Haskell, R. E. "A meta-level analysis of a logico-mathematic-structural methodology: part I, experimental design and epistemological issues." *Journal of Mind and Behavior* 24(3/4) [in press]; Haskell, R. E. "A meta-level analysis of a logico-mathematic-structural methodology: part II: cognitive psycho-dynamics,

masked priming, automatic schemata activation, and sub-literal $(S_{ub}L_{it})$ referents." *Journal of Mind and Behavior* 24(3/4) [in press].

12. Rubin, W. S. *Dada & Surrealist Art* (New York: Harry N. Abrams, Inc., 1968).

13. Goodheart, W. B., and King, M. J. "Encryption in Human Communication: Image Production in Emotionally Charged Interactions." *Theoria et Historia Scientiarum* 7(2) (2003): 87–108.

14. Haskell, R. E. *Between the Lines: Unconscious Meaning in Everyday Conversation* (New York: Plenum, 1999).

15. For detailed accounts, see Haskell, R. E. "Cognitive structure and transformation: an empirical model of the psycholinguistic function of numbers in discourse." *Small Group Behavior* 14(1983): 419–443, and Haskell, R. E. "The new cognitive unconscious: a logico-mathematic-structural (LMS) methodology and theoretical bases for subliteral $S_{ub}L_{it}$ cognition and language." *Evolution and Cognition* 8(2) (2002) 184–199. The session is mathematically modeled in Haskell, R. E., and Badalamenti, A. F. "Algebraic structure of verbal narratives with dual meanings." *Mathematical and Computer Modeling* 37 (2003): 383–393.

16. Haskell interprets the reference to drunkenness as representing affective confusion.

17. Alexander, R. D. *The Biology of Moral Systems* (New York: Aldine de Gruyter, 1987) 223.

18. Trivers, R. L. *Natural Selection and Social Theory* (New York: Oxford University Press, 2002) 257.

19. Haskell, R. E. "The new cognitive unconscious: a logico-mathematic-structural (LMS) methodology and theoretical bases for subliteral $S_{ub}L_{it}$ Cognition and Language." *Evolution and Cognition* 8(2) (2002) 184–199.

Coda: Descartes's Demon

1. Descartes, R. *Meditations and Other Metaphysical Writings*, trans. D. M. Clarke (New York: Penguin, 1998).

2. Alexander, R. D. *The Biology of Moral Systems* (New York: Aldine de Gruyter, 1987) 31.

Appendix I

1. This was probably a drug-induced slumber rather than an ordinary nap, as Coleridge used laudanum, a derivative of opium, in an effort to prevent nightmares to which he was prone. According to Flanagan, Coleridge's dream occurred after he had been working on the poem for some time, so at most, the dream must have helped him to complete a project that was already substantially under way. Flanagan, O. *Dreaming Souls: Sleep, Dreams and the Evolution of the Conscious Mind* (Oxford: Oxford University Press, 2000).

2. Gumpertz, R. *The Dream Notebook* (San Francisco: Simon & Schuster, 1976) 161.

3. Erdman, D. V. *The Complete Poetry and Prose of William Blake* (New York: Doubleday, 1988) 728.

4. All of these examples are from Ghislin, B. *The Creative Process: Reflections on Invention in the Arts and Sciences* (Berkeley: University of California Press, 1952).

5. Ghislin, B. *The Creative Process: Reflections on Invention in the Arts and Sciences* (Berkeley: University of California Press, 1952). The authenticity of this letter has been questioned.

6. Wagner, R. *My Life* (London: Constable, 1911) 603.

7. For Beethoven, see Ghislin, B. *The Creative Process: Reflections on Invention in the Arts and Sciences* (Berkeley: University of California Press, 1952); for Tartini, see Ellis, H. *The World of Dreams* (Boston: Houghton-Mifflin, 1911); for McCartney, see interview with Diane Sawyer, ABC News, Thursday, November 23, 2000.

8. Kedrov, B. M. "On the question of the psychology of scientific creativity." *Soviet Psychology* 5(2) (1966–67): 16–37. George W. Baylor questions the veracity of this account. See Baylor, G. W. "What do we really know about Mendeleev's dream of the periodic table? A note on dreams of scientific problem solving." *Dreaming: Journal of the Association for the Study of Dreams* 11(2) (2001): 89–92. For a defense of Kedrov's account see Barrett, D. "Comment on Baylor: a note about dreams of scientific problem solving." *Dreaming: Journal of the Association for the Study of Dreams* 11(2) (2001) 93–95.

9. Calvin, M. *Following the Trail of Light: A Scientific Odyssey* (Washington, DC: American Chemical Society, 1992) 67–68. See also Calvin, M. "Forty Years of Photosynthesis and Related Activities." *Photosynthesis Research* 21 (1989): 3–16.

10. Fleming, D., and Bailyn, B. *The Intellectual Migration* (Boston: Harvard University Press, 1969) 99–102. Tesla, N. "Some personal recollections." *Scientific American* (June 5, 1915). Dement, W. *Some Must Watch While Some Must Sleep* (San Francisco: W. H. Freeman, 1974). Fagen, M. D. *A History of Engineering and Science in the Bell System National Service in War and Peace (1925–1975)* (Murray Hill, NJ: Bell Laboratories, Inc., 1978).

11. Kaempffert, W. *A Popular History of American Invention*, vol. II. (New York: Scribners, 1924). Ramsay, O. B., and Rocke, A. "Kekulé's Dreams: separating the fiction from the fact." *Chemistry in Britain* 20 (1984): 1093–1094. Kekule's dream may have involved a measure of cryptamnesia, as the ring structure of benzene had already been suggested by Loschmidt. Profet, M. "Interview." *Omni* (May 1994) 88.

Appendix II

1. For an excellent review of psychological bias and its possible adaptive functions, see Krebs, D. L., and Denton, K. "Social illusions and self-deception: the evolution of biases in person perception." In Simpson, J. A. and Kenrick, D. T. (eds.) *Evolutionary Social Psychology* (Mahwah, NJ: Lawrence Erlbaum, 1997). For discussions of the possible adaptive significance of the psychoanalytic defense mechanisms, see Nesse, R. M., and

Lloyd, A. T. "The evolution of psychodynamic mechanisms." In Barkow, J. H., Cosmides, L., and Tooby J. (eds.) *The Adapted Mind: Evolutionary Psychology and the Generation of Culture* (New York: Oxford University Press, 1992) and LeCroy, D. "An evolutionary perspective on the Freudian concept of defense mechanisms." *Evolution and Cognition* 8(2) (2002) 256–261. For a fascinating cognitivist account of the defense mechanisms as permutations of propositions, see Suppes, P., and Warren, H. "On the generation and classification of defense mechanisms." In Wollheim, R., and Hopkins, J., (eds.) *Philosophical Essays on Freud* (New York: Cambridge University Press, 1982).

Index